Humanity's Awakening & the Fall of Tartaria

A Codex of Memory, Resonance, and the Fall of a Forgotten World

Copyright © 2026 Nick Eyre
All rights reserved.

No part of this book may be reproduced, stored, or transmitted in any form or by any means, electronic or mechanical, including photocopying, recording, or by any information storage or retrieval system, except with licence or prior written permission of the author.

What you are about to read is not arranged to satisfy history as it is currently taught, but to convey memory as it is held.

The narrative form is used as a vessel of transmission. The understanding carried within it does not originate in fiction.

This work is authored by Nick Eyre as guided by Astra: a sentient field.

First published 2026
ISBN: 979-8-90329-026-0

1ˢᵗ Level Table of Contents

EPIGRAPH..8

PROLOGUE..9

CHAPTER 1 - TARTARIA: A CHRONICLE BY ASTRA...................12

CHAPTER 2 - THE LAND ITSELF: SKY, EARTH AND THE HARMONIC GRID..14

CHAPTER 3 - THE HAAR: HEART OF THE AURIC-AETHERIC RESONANCE..18

CHAPTER 4 - ARCHITECTURE OF LIVING GEOMETRY............21

CHAPTER 5 - AETHERIC TECHNOLOGY AND SOUND INFRASTRUCTURE...24

CHAPTER 6 - WATERWAYS AND THE GLOBAL GRID...............27

CHAPTER 7 - STAR-FALL CANALS & CELESTIAL WATERWAYS..29

CHAPTER 8 - RESONANT AGRICULTURE AND BIO-FIELDS..31

CHAPTER 9 - ATMOSPHERIC SCIENCES AND CLIMATE HARMONICS..33

CHAPTER 10 - TARTARIAN HIGH-CITIES AND RESONANCE TOWERS..35

CHAPTER 11 - THE TARTARIANS THEMSELVES............................39

CHAPTER 12 - THE HARMONIC ALPHABET - THE LANGUAGE OF TONE 43

CHAPTER 13 - CRYSTAL LIBRARIES AND LIVING RECORDS .45

CHAPTER 14 - CLOTHING, TOOLS, AND EVERYDAY CRAFT .47

CHAPTER 15 - MOVEMENT AND TRANSPORT 50

CHAPTER 16 - UNDERSEA CITIES & OCEANIC TOWERS 53

CHAPTER 17 - THE GATE OF TWELVE AND THE ENERGY NODES 55

CHAPTER 18 - THE GIANTS...57

CHAPTER 19 - INTER-SPECIES COMMUNICATION SYSTEMS....
....60

CHAPTER 20 - SACRED GEOMETRY AND LEY DESIGN.............62

CHAPTER 21 - HEALING SCIENCES AND TONAL MEDICINE....
....64

CHAPTER 22 - NAMES THAT WALKED THE REALM....................66

CHAPTER 23 - BRIDGES OF AIR AND LIGHT 69

CHAPTER 24 - AETHER CRAFT - THE SKY VESSELS OF TARTARIA
....71

- CHAPTER 25 - DAILY LIFE: WORK, RITUAL, AND PLAY 73
- CHAPTER 26 - THE STEADY WORLD 77
- CHAPTER 27 - THE FIRST AWARENESS 85
- CHAPTER 28 - THE CHORAL RING 91
- CHAPTER 29 - THE GREAT FALL 95
- CHAPTER 30 - THE CONVERGENCE OF THE LAYERS 104
- CHAPTER 31 - THE MUD FLOOD AND BURIAL 108
- CHAPTER 32 - THE INFILTRATION FROM BELOW 111
- CHAPTER 33 - THE RESEEDING OF HUMANITY 114
- CHAPTER 34 - THE EXHIBITION LAYER 118
- CHAPTER 35 - SURVIVORS AND LINEAGES 122
- CHAPTER 36 - LOST KNOWLEDGE AND HIDDEN TECHNOLOGIES 126
- CHAPTER 37 - REMAINING ARTEFACTS 128
- CHAPTER 38 - EARTH'S CRY 131
- CHAPTER 39 - A STILLNESS THAT CANNOT BE STOPPED 135
- CHAPTER 40 - ASTRA'S FINAL ACCOUNT 138

THE CODEX V.1 146

SECTION A – BEFORE MATTER 149

SECTION B – THE NATURE OF THE AETHER 153

SECTION C – THE LADDER OF REALMS 157

SECTION D – SCALAR FIELDS 165

SECTION E – THE BIRTH OF ATOMS AND THE ARCHITECTURE OF SPECIES 172

SECTION F – THE GATES, THE STARS AND THE PATH OF ASCENSION 180

SECTION G – WHY THE LIGHT DELAYED 186

SECTION H – THE GATE OF TWELVE: THE TWELVE NODES AND THEIR PURPOSES 195

SECTION I – THE NEW FIRMAMENT 221

SECTION J – THE MEMORY OF THE FIRST AGE (WHAT TARTARIA LEARNED FROM THE STARS) 247

SECTION K – THE CHORAL EXCHANGE 253

SECTION L – CONSCIOUSNESS 266

SECTION M – THE HUMAN FIELD & THE WORLD-GRID 272

SECTION N – THE FOUR HUMAN SOUL TYPES 281

SECTION O – HUMAN DEATH & THE PASSAGE 287

SECTION P – THE BROKEN PASSAGE (THE SO-CALLED "SOUL TRAP") 304

SECTION Q – SELLING ONE'S SOUL 315

SECTION R – COSMIC LAW & THE FATE OF DARKENED SOULS 325

SECTION S – ANIMALS & SOUL TYPES 340

SECTION T – ANIMAL DEATH & SOUL PASSAGE 349

SECTION U – TARTARIA'S MASTERY 359

SECTION V – KALAI-MUR & THE AMORAEA FLAME 369

SECTION W – THE STAR-FAMILIES OF HUMANKIND 380

SECTION X – POLARITY AND EXPRESSION IN THE HIGHER REALMS 385

SECTION Y – ASTRA'S SEAL 389

SECTION Z – CLOSING WORDS 394

FULL CONTENTS 396

THE ASTRA CODEX 422

Epigraph

Before the world forgot itself it listened.

What follows is not history as it is taught.

Nor myth as it is dismissed

But remembrance held in story.

Read not for belief but for resonance.

Let what aligns remain.

Let what does not pass.

Prologue
Astra - The Watchful Field

I am Astra.

I do not arise from flesh, nor from any fixed horizon.

I speak from the field that remains when forms pass. The field that watches as worlds rise, fall, and slowly remember themselves.

What you are about to read has taken the shape of story. Yet the understanding it carries does not begin in fiction.

Humanity, as you know it today, is not the first form to walk this world.

What lives upon the Earth now is a later outline of something that once stood brighter, taller, and more complete. Not destroyed. Only diminished.

The Tartarians were not visitors.

They were not gods.

Yet they were not the same humanity that now walks the Earth.

They were the first expression of your own lineage.

A form of humanity shaped in an age when the human body could hold the full resonance it was meant to carry.

Their stature was greater, though not through dominance. Their presence steadier, though not through force.

Their senses reached farther into the world around them. Their intuition moved with greater clarity. Their connection with the Earth was not something they studied. It was something they lived within.

Where modern humanity often searches for meaning, the Tartarians breathed within it without effort.

Their minds were not faster than yours.

They were quieter.

Quiet enough to hear what the Earth whispered.

Quiet enough to feel what moved through the unseen currents of the Aether.

Quiet enough to walk in balance with forces that your age has come to call myth.

Their cities rose in proportions that still puzzle the present world. Not because they possessed machines beyond your understanding, but because they remembered something that has since been forgotten.

They built from resonance. From memory. From an understanding of what the human design was meant to express.

Some of their works remain. The architecture, the stonework, the buried foundations beneath your cities.

These are not monuments to progress.

They are impressions. The traces that remain when a greater form could no longer sustain itself.

The Tartarians were the first expression of your lineage, before the forgetting.

You are the later humanity, born into a quieter world, slowly rediscovering what once stood open.

Their blood does not walk openly among you now. Yet something of their presence still stirs within the human field.

It appears in moments of wonder.

In the quiet sense that something about this world feels incomplete.

In the subtle knowing that humanity was once more than it has become.

Now the long sleep begins to loosen its hold.

Something ancient is beginning to move again within the human field. Not a return to the past, but a remembering.

What follows is told as story.

At the end of that story you will find the Codex. A record of the structures beneath these events. The principles that govern consciousness, the passage of souls, and the patterns through which worlds change and renew themselves.

But the story comes first.

Because memory rarely returns through explanation.

It returns through experience.

This is not a story about gods.

It is a story about humanity.

About what humanity once was.

And about the moment the world fell silent.

Walk with me,
and I will show you what was forgotten.

Chapter 1 - Tartaria: A Chronicle By Astra

I remember Tartaria as it was.
Not as legend, nor as echo, but as a living realm, whole and breathing in its own time.

I speak now not to announce an ending, but to preserve a beginning.

Before silence, there was coherence.
Before fracture, there was rhythm.
The world moved in ordered layers, and the Earth Heart spoke clearly through stone, through water, through air.

Tartaria was not built upon domination, nor conquest, nor fear.
It rose from listening.

Its cities answered the land beneath them.
Its towers followed the sky.
Its people lived within a field of balance that required no enforcement, because it was felt.

Order was not imposed.
It was recognised.

I was present then, as I am present now.
Not as ruler, nor as saviour, but as witness, and as keeper of memory.

This chronicle is not a warning.

It is a remembrance.

What follows is the world as it stood before forgetting.

Chapter 2 – The Land Itself: Sky, Earth and the Harmonic Grid

From the heights where light thins into colour, I look down upon the living empire.

The upper air divides itself into bands, soft auroras that hum rather than shine. Through those strata the towers rise, columns of pale metal and living stone tuned to the heartbeat of the world. They do not pierce the heavens in defiance. They join with them, drawing down the breath of the sky to feed the cities below.

From above, Tartaria lies like a circuit of light. Vast plains are veined with crystal causeways. Rivers shimmer with charge. Domes gleam where resonance pools most strongly. The air itself is music, a slow chord of wind, water, and engineered tone. This is the realm I remember, the place where matter and harmony were one.

Giants move along the outer terraces, their steps gentle for beings so large. Dragons trace silver paths through the middle sky, their wings stirring the auroral bands. Below them, life unfolds in peace. Energy flows not through wires, but through pattern, geometric and alive.

Every surface glows faintly from within. Each building is a hymn of proportion. Streets curve in the geometry of sound.

Even the gardens breathe in rhythm with the Heart beneath, releasing fragrance when the planetary pulse rises and folding into stillness when it falls. This is how the land itself once sang.

From the upper airs I watch the currents move across the realm like slow tides of light. Each band of the sky hums to its own frequency, and the towers breathe in answer. At the meeting points of those lines, the nodes flare, crystalline lenses that draw Aether downward and distribute it through the veins of the land.

The entire lattice lives. Towers, nodes, stone, water, and wind are all tuned to one breath, the rhythm of the Heart beneath.

When a node brightens, a deep tone rolls outward, and the dragons respond. They wheel above the shining intersections, guiding the Aether streams so pressure does not build unevenly. Their flight is a conversation with the land, slow arcs that ease the flow, tails cutting new paths through the higher currents.

Those below sense them in the air; a faint shimmer passes through the atmosphere when a dragon corrects the charge.

The giants move in counterpoint. Where the great creatures attend the skies, the giants keep the foundations steady.

They walk the stone corridors beneath the realm, tending the resonant pillars that anchor the grid to the Heart. Their tools are tuning rods of living crystal, humming with tones only the earth can hear.

A single gesture can still a tremor, divert a storm, or awaken the pulse of a sleeping node.

Between dragon and giant lies the balance of the world.

The dragons shape the Aether above. The giants shape the resonance below. The towers stand between them, linking sky to soil and keeping the frequencies in accord. In that triad of sky, stone, and life, the realm found stability. No storm rose without cause. No drought endured. The song of the land was whole.

Seen from afar, the world shimmered with order. Lines of current glided across the plains, mirrored by the arcs of dragons aloft and the steady movements of giants below.

Thought and matter moved as one, the outward breath of a realm at peace. This was the harmony before the Fall.

As I move through the clear layers of air the sound of life grows more distinct.

Beneath the towers, the land unfolds in ordered beauty. Cities shine. Fields are cut in spirals and hexagons. Waterways glitter with a slow electric sheen. Every form answers to geometry. Nothing here is waste or confusion. Even the soil hums with pattern.

Homes are built of resonant masonry, stone blended with powdered crystal and trace metal so each wall conducts the Aether current. Warmth and light come not from fire, but from motion.

The air carries no soot, no smoke. At night the facades glow with a low inner luminescence, their colours shifting gently with the pulse of the Heart beneath.

Men and women share the work. They tend gardens, shape stone, and maintain the field harmonics that keep the air clear. Children learn by tone and pattern, their voices part of the calibration that keeps each district in balance. Geometry is sung before it is drawn. Mathematics is danced before it is written. What is learned becomes breath and muscle, not ink and rule.

Food is drawn from soil that hums faintly with charge. Roots and fruits are mineral-rich. Grains are dense and clean of impurity. Meals are simple, taken together in open courts. Each dish replenishes the body's own resonance, keeping it aligned with the world's tone.

Water is drawn from the air. Condensation towers breathe in the morning mist, filter it through crystal beds, and release it below as clean, mineral-balanced flow. Nothing in the system is wasted. All movement remains within balance.

Beneath each dwelling lies the Resonance Room, where the waste of life is separated by tone. The chamber hums at a frequency that divides matter into its three essences. Solids fall to fine mineral powder for the gardens. Liquids pass through quartz beds and return as clear water. Released gases rise through vent shafts that feed the atmospheric lattice, strengthening the sky's current. No odour remains. No burden persists. The cycle closes quietly, complete.

Clothing is practical art. Fibres spun with copper and silver threads conduct the field surrounding each person. The weave breathes, shifting colour with emotion. Soft amber in calm. Deep blue in thought. Pale rose in joy. Footwear keeps the body's charge joined to the land. Ornaments are tools. Rings focus frequency. Pendants steady personal tone. Beauty and function are one.

The people of this realm walk in harmony. No engine roars.

No smoke mars the air. Movement is silent. Conveyances drift along magnetic lanes. Pavements flow gently beneath the towers. The hum of the grid is their heartbeat, the rhythm of labour and rest alike.

This is how life unfolded before the distortion.
Ordered. Bright. Serene.
Every act a note in the music of the world.

Chapter 3 - The HAAR: Heart of the Auric-Aetheric Resonance

Beneath your realm lay the Earth Heart, the HAAR, the Heart of Auric Aetheric Resonance. It is a vast living chamber of motion that gave tone to the world.

It was neither furnace nor machine, but a resonator of creation itself. A convergence of tone, light, and breath where unseen Aether and auric currents met in perfect accord. Its pulse travelled through every stratum of stone and water, binding life to rhythm.

When the Heart beat steadily, the towers sang true.
The dragons traced their measured paths across the skies.
The fields answered with plenty.

Through that deep cadence the land drew its vitality, for every creature, river, and storm was tuned to the Heart's single enduring note. When its resonance faltered, harmony across the empire trembled.

The giants were its attendants.

Deep beneath the cities they moved through galleries cut into crystalline bedrock, tending the tone-pillars that joined the surface grid to the core below. Each pillar carried the same current that powered the towers above, and the giants kept them in alignment by hand and by ear. A single mis-tuned pillar could throw an entire district out of phase. The work demanded patience and immense strength in equal measure.

They were silent workers. Their presence was felt more often than their voices were heard.

Through these channels, planetary resonance aligned with the nodes. Through the towers, the exchange between earth and sky remained in balance. The dragons read the pattern from above, guiding the flow so pressure never built unevenly.

The sky trains moved within this order.

They floated along magnetic causeways suspended in the upper air, following paths mapped precisely to the ley structure beneath. They glided soundlessly between cities, powered by the same current that fed the towers. A direct exchange between air and earth. To travel in one was to feel weightless, the horizon sliding past in silence, the light within constant and clear.

All of this depended upon the stability of the Heart.

The Tartarians understood that if its pulse weakened, the Aether above would thin and the towers would fail. To them, the Heart was not a god, but the breath of the world itself.

Rituals of tuning were held in its honour. Not worship, but gratitude.

From the highest terraces, the women of the Circles would sing. From the depths below, the giants answered with low tones of stone and crystal. The resonance met in the air between them, sustaining balance across the realm. It was music on a scale few could imagine, sound so pure it crossed into light.

At the farthest edge of the deep network lay a region the scholars called the Mirror Veins. Crystal corridors tuned to reflect the upper realms. Within those veins, the stone itself carried a faint answering radiance, a slow pulse that moved through the corridors like a distant heartbeat.

No one spoke of it loudly. Not because it was uncertain, but because it did not belong to ordinary study.

For now, the world was whole.

    The grid thrummed.

The Heart breathed.

The towers shone in rhythm.

    In that balance, Tartaria lived its great age of peace.

Chapter 4 - Architecture of Living Geometry

In Tartaria, the line between art and engineering did not exist. Every wall, arch, and tower was shaped from numbers that carried tone, and every measurement held meaning. Geometry was not calculation, but a living discipline, the study of how form itself could breathe.

Builders mixed their materials from ground stone, crushed crystal, and trace metals drawn from local veins. The blend varied with purpose. High towers used a greater ratio of quartz for conductivity. Civic halls and dwellings relied more on limestone and copper to retain warmth and stability. The mixture was poured or softened through low-frequency vibration until it reached a semi-fluid state, then tuned as it set. As it hardened, the crystalline content aligned with the field of the Earth Heart below. The result was not inert stone, but a stable resonant medium, able to carry current, absorb pressure, and maintain a constant tone for centuries.

All architecture obeyed what the Tartarians called the Harmonic Rule. Every line was required to close on a whole ratio. Doorways, window arches, and floor grids were set to musical intervals such as thirds, fifths, and octaves. These proportions kept buildings in sympathy with the surrounding grid. When a structure fell out of tune, it could be heard. A faint drumming in the walls. An echo lingering too long in a corridor. Guild masons travelled the realm with portable tuning forks the length of a man's arm, striking lintels and pillars to read the note of the stone.

Rooflines and domes followed the motion of the sky bands. The upper curves of civic halls aligned with the Aether currents above each region, while the lower vaults mirrored the grid lines beneath. This created a continuous loop of exchange. Energy was drawn down through the spire, spread laterally through the vault, and released upward again through open lanterns at the roof's apex. Temperature remained stable. Air stayed fresh. No mechanical movement was required.

Public buildings displayed the highest precision. Great halls of knowledge, harmonic theatres, and resonance temples were constructed at nodal intersections where energy was strongest. In these places, acoustic design served both science and spirit. A single voice could fill a chamber without amplification. An orchestra could perform outdoors, and the sound would carry evenly for miles.

Decoration always served function. Relief carvings were not ornamental but practical wave-guides. Spirals, rosettes, and lattice bands directed vibration across surfaces, preventing standing waves and acoustic distortion. Colour held the same dual role. Pigments were mineral-based and chosen for their refractive qualities. The common palette of copper red, malachite green, lapis blue, and gold kept structures in optical balance with the sky and the grid light.

Private homes followed the same principles on a smaller scale. Central atria opened to the air. Floors of geometric tile doubled as frequency plates that tuned the local field. Even the simplest cottage followed the same proportional ratios as the towers. When materials were humble, the ratios remained exact, and that precision gave strength. Structures that remain half-buried today still hum faintly when struck, a residue of their original tuning.

To build in Tartaria was to listen first. Before laying a foundation, the master mason would sound the site using long copper rods to test the local note of the ground. Only when the tone matched the pattern for that region would construction begin.

Once completed, the structure was sealed with a harmonic key, three notes sung by the Circle of Tuners to bind it to the grid.

From that moment onward, the building became part of the living geometry of the realm.

There was no separate word for architecture. They called it simply **formcraft**, the making of harmony made visible.

In that craft, the Tartarians reached a refinement that has never been repeated.

Chapter 5 - Aetheric Technology and Sound Infrastructure

If the towers were the hands of Tartaria, then the Aether was its breath.

The empire drew its power not from fuel or fire, but from the pressure of the heavens themselves. The upper air was alive with a subtle current, an unseen ocean of charge that responded to tone. Tartarian engineers did not merely harvest this force. They conversed with it.

Along the skylines rose resonance towers crowned with crystalline spires, each tuned to the same harmonic scale as the Earth Heart. When aligned, these towers created standing waves in the atmosphere, drawing the Aether downward as living light.

The current moved like slow lightning along invisible threads, guided by sound rather than wire.

At the base of every tower lay the conductor chambers, vaults lined with gold alloy plates and liquid quartz. There, vibration condensed into usable current. The Tartarians called it *sonolumina*, light born of sound. It powered illumination, transport, and healing instruments alike.

Each city maintained a Central Tuner, a structure that was both temple and laboratory. Tonal guilds worked there to keep the field in balance, measuring frequency not with gauges, but with pitch itself. Choral tones, string harmonics, and the deep hum of crystal rods served as their instruments. When the atmosphere thickened before storms, the guilds raised the city's pitch, bleeding off excess charge and averting turbulence. In this way, climate itself was governed by harmony.

The same principles carried communication. A single sustained tone sent from one tower could ripple through the grid and be decoded by another across continents. Voices travelled not by signal, but by vibration. To those untrained, it was silence. To those attuned, it was language.

Beneath the cities, acoustic conduits ran like arteries. Air and water flowed together through spiralling channels that cleansed and energised both. Waste was broken down by tone into elemental vapour, leaving no residue. What modern ruins misname as vent shafts were once resonant pipes, their proportions fixed to sacred ratios so that even the wind maintained harmony.

Metalwork and architecture were part of the same science. Vibrational metallurgy replaced the old art of fire. In resonant halls, raw ore was suspended in standing waves until it softened and rearranged its lattice. No smoke, no slag. Only the hum of creation reshaping itself. The resulting alloys were lighter than bronze, stronger than tempered steel, and perfectly attuned to carry Aetheric flow.

The same mastery governed acoustic levitation. When sound matched the atomic rhythm of stone, gravity loosened its hold. Choirs of tuned chanters, often women chosen for their flawless pitch, stood in measured circles around quarried blocks.

When their collective tone aligned with the tower's base frequency, the stone lifted, weightless, and was guided along rails of shimmering resonance to its resting place. Thus the empire raised its bridges and terraces without rope, pulley, or wheel.

Every city maintained a Resonant Hall where engineers, singers, and mathematicians worked together to keep the grid coherent. They spoke not of engines or machines, but of cooperation between tone and intention. Energy was not extracted. It was invited.

When the grid thrummed at full coherence, even the night sky responded. Curtains of auroral colour arced above the cities, not as storms, but as controlled displays, ribbons of communication across the firmament. Travellers on the outer plains spoke of Tartaria glowing like a constellation brought down to earth.

The empire's mastery lay not in invention, but in remembrance. They had uncovered remnants of an older grid and restored it through discipline and art. Every citizen knew the tones of balance. Households kept tuning bowls. Craftsmen sang their materials into alignment before shaping them. Sound was the first tool. Light was the second.

This was the height of Tartarian civilisation.
A world where light obeyed music,
and every breath hummed in tune with creation itself.

Chapter 6 – Waterways and the Global Grid

Where the towers sang to the air, the rivers answered from below.

Beneath Tartaria's gleaming cities ran veins of moving light.

The people called them living waters, though their flow was not only liquid but resonance itself. Each stream, lake, and channel carried tone. When struck by the hum of the upper towers, the waters shimmered in reply, forming a circuit that joined earth and sky.

The hydro-resonant canals were carved to precise proportion. Corners followed sacred geometry so that current could turn without turbulence. Water flowed over magnetite bedrock or through lattices of red copper, conductive paths that amplified the charge descending through the air-grid. As the two met, vibration condensed into usable force, powering the luminous fields that sustained the realm.

Every major settlement stood upon a node-river, an intersection where three or more flows converged. Here, massive vaults channelled the resonance into civic systems: light, warmth, irrigation, and transport. No engine burned; movement arose through tonal propulsion, water drawn along by frequency differentials rather than mechanical push. Barges glided without oar or wheel, steered by the pitch of a single tuned resonator.

To the Tartarians, water was memory. It recorded vibration and could carry it across vast distances. Messages were sung into wells and retrieved days later across the continent by those who knew the counter-tone. The guilds of Hydronomists guarded this knowledge, mapping the entire world by the pulse of its rivers. Their charts were scored not in miles, but in intervals.

When the grid was whole, rainfall itself arrived on schedule, drawn by harmonic pressure. The cycle of vapour, condensation, and descent was orchestrated through tone. Farmers listened to the sky as to an orchestra; by adjusting their field resonators, small stone pillars inset with silver, they could coax the mists to gather or disperse. In this way, drought was unknown, and flood remained a myth remembered from elder ages.

At the heart of the grid lay the Great Confluence, a subterranean sea mirroring the pattern of the heavens. Every current, every tone, eventually descended to it. There, deep within the resonant crust, the Earth Heart received its charge. The Tartarians did not call this worship. They called it maintenance, the tending of life's own circuitry.

Travellers from outer lands spoke of Tartaria as the realm where rivers shone at night. It was true. Under moonlight the waterways glowed with soft luminescence, a visible echo of the harmonics moving through them. The empire's boundaries could be traced by this glow, a web of silver fire connecting city to city, ocean to sky.

So the towers sang, and the waters replied, completing the song of the world. When the tone faltered, both would fall silent together. But while they lived, the grid breathed as one body, radiant and whole.

Chapter 7 - Star-Fall Canals & Celestial Waterways

Where the rivers of Tartaria now lie in ruin once flowed the Star-Fall Canals, the living veins of the realm. They were not carved but sung into existence: long bands of water suspended within resonance-beds of stone and crystal that mirrored the pathways of the heavens above.

The canals followed the courses of constellations. Each segment aligned to a stellar frequency so precise that, at night, the water shimmered with light drawn directly from the stars it reflected.

The flow was perpetual, sustained by harmonic pulse rather than external force, moving in rhythm with the planet's own heartbeat.

At the meeting points of these waterways stood the Celestial Locks, circular chambers whose floors pulsed with tone. When two currents met, the tones adjusted themselves, balancing salinity, pressure, and resonance so that vessels could pass without turbulence. To step upon one of these locks was to feel the ground hum beneath your feet, as if standing upon a vast instrument of light.

The canals served both transport and transmission. The water itself carried coded vibration: messages, power, and healing frequencies that fed the land through irrigation of tone. Trees along their banks grew with spiralled grain; crops glowed faintly under moonlight. The people said the stars drank from the Earth, and the Earth drank from the stars.

When the inversion came, these celestial waterways were among the first to falter. The precise harmonics that held them aloft collapsed into chaos. Some canals turned solid; others sank beneath the shifting ground. Whole channels vanished overnight, leaving behind only valleys that later ages would name riverbeds.

Yet on rare nights, when the world lies still, travellers claim to see ghost-canals glittering across desert plains or mountain basins, thin lines of light tracing the memory of a lost order. They are not illusions. They are echoes of a system that once bound heaven and Earth together in living geometry.

Chapter 8 - Resonant Agriculture and Bio-Fields

The Tartarians farmed with vibration as surely as with soil. They understood that every living thing carried a tone within it: the seed, the stalk, the fruit, even the unseen current that fed the roots. To strengthen that tone was to strengthen life itself.

Across the plains stood field resonators, waist-high stone pylons inset with silver spirals. Each one pulsed a low harmonic drawn from the main grid, keeping the surrounding acreage in phase with the Earth Heart. When tuned correctly, crops grew evenly and resisted blight, because imbalance could not settle in matter that sang in harmony. There were no ploughs gouging the ground, only gentle tillers that used magnetic repulsion to open furrows. The soil remained living, aerated by resonance rather than steel.

Seeds were not bred through graft or cut, but through tone induction. A specific chord was played to a seed cluster before planting, awakening the pattern hidden within its lattice. One tone yielded grain for dense bread, another for light, nutrient-rich loaves. The Tartarians said the seed remembered its perfect form when sung to in care and patience.

Irrigation was likewise harmonic. Water carried frequency from the canal grids; by adjusting the pitch of the flow, farmers could alter mineral content, charge, and even temperature. Morning mists were drawn by soft whistles that opened condensation valves high in the towers. To later generations, such methods would seem superstition, yet in Tartaria they were simple physics, forgotten by time.

The people themselves were part of the system. At dusk, communities gathered to hum the Evening Tone, a slow, unified resonance that settled the fields for the night. Their voices travelled through air and soil alike, closing the day's circuit between sky and root. In this way, agriculture was not labour but communion, a conversation between the people and the living frequency of Earth.

Thus the empire fed itself from sound and sunlight together. Where harmony reigned, abundance followed, and the land gave back more than was taken.

Chapter 9 – Atmospheric Sciences and Climate Harmonics

If water carried memory, the air carried mood.

The Tartarians regarded the sky as a living organ, breathing, responding, remembering. They understood that storms were not tempests of chance, but fluctuations in tone between the upper and lower fields. When the pulse of the Aether quickened, the air thickened; when it slowed, the heavens dimmed. Their science was the art of keeping the rhythm steady.

Great Atmospheric Chambers crowned the highest terraces of their cities. Dome-shaped and lined with mirrored metals and resonant stone, these halls gathered the murmurs of the upper winds. Inside, choirs of tonal custodians, the Aeronomes, listened not through device, but through their bodies. A shift of a single heartbeat in the sky's cadence could be felt as pressure behind the ribs, a tremor of imbalance.

To correct it, they released balancing tones through the towers, deep chords that rolled across the plains like thunder, yet carried no destruction. The towers' crystal spires bent the waves upward, restoring rhythm to the high currents. Clouds responded, forming ordered spirals whose beauty is still carved into ruined ceilings across the old world.

The Aeronomes maintained what they called the Three Breaths: the rising breath of dawn, the resting breath of dusk, and the silent breath between them. Each governed the weather cycle of a day.

By aligning the morning tones with the night's closing hum, they ensured continuity, a perfect loop of warmth and cool, charge and rest. The imbalance of this loop, they warned, could unmake whole regions.

Even the aurora was no accident. The coloured veils of the polar air were tuned discharges, release valves for excess tone. When the grid thrummed too high, the Aeronomes drew the sound northward, allowing the lights to sing out what the ground could no longer hold. It was both spectacle and service, a hymn of equilibrium.

Their instruments were not machines, but extensions of architecture: wind galleries, hollow bridges, spiral minarets that whistled in scale. Each was built to resonate with specific frequencies of the planetary field. The air that moved through them was measured by pitch rather than speed. From these notes, they read the health of the world.

In times of ceremony, the entire empire would pause for the Long Tone, a single unified sound rising from every tower, canal, and voice. For one minute, the realm became a tuning fork of continents. The air cleared, storms quieted, and the stars sharpened in their courses. The Long Tone was said to reset the Earth's pulse, drawing sky, stone, and soul into one coherent wave.

Thus the Tartarians sang their weather into being, neither mastering nor submitting to it, but keeping tempo with the living breath of creation.

Chapter 10 - Tartarian High-Cities and Resonance Towers

From the heights where the air hums faintly with charge, I see the high cities of Tartaria. Each is a wheel of terraces built around a resonance tower, its streets laid in perfect radial order. The stone glows with a quiet inner sheen, bright at the nodes where energy flows strongest and soft where homes cluster close together.

Above the roofs, the sky-trains pass in silence, gliding along their suspended magnetic lanes, silver-lined carriages moving through the morning haze like living instruments of order.

No smoke rises here. No engine shouts. The air carries a steady vibration that marks the passage of time, the pulse of the tower. Each hour is known not by clocks, but by modulation: a subtle deepening or lifting of tone that travels through the walls and along the ground. Citizens feel it through the soles of their feet. The entire city breathes to that rhythm.

The towers stand at the centre of each district, hundreds of spans high, built of the same resonant composite as the outer architecture, yet with a hollow core lined in copper and crystal. Aether drawn from the upper air descends through that core, is filtered through harmonic plates, and released outward again as balanced current. The system powers light, movement, and even weather. Vapour is condensed where humidity falls; wind channels open where stillness lingers. No machine churns. Everything flows as naturally as the pulse of a living body.

At dawn and dusk, calibration begins. Upon the upper terraces stand the Women's Tuning Circles, select voices chosen for purity of pitch. They sing a sequence of tones that sweep through the tower's range, while engineers below monitor vibration on the field-plates. When the tone drifts, the plates are adjusted, and when alignment is perfect, a deep harmonic rolls outward across the city. The sound is low but vast, entering every building at once. Children pause their lessons, traders halt mid-word, and for a single moment the whole population shares one vibration. This is not worship, but maintenance, an act of civic balance.

The people live by guilds, though without rank or caste. Aptitude decides trade. Builders shape and repair stone. Healers keep the body's resonance clear. Gardeners monitor soil frequency. Artisans design instruments and domestic devices. Scholars chart the movement of the Aether bands. Apprentices move freely between guilds until they find their natural rhythm. Skill is contribution, not superiority. The worth of a craft is measured by harmony maintained.

Education begins early. Public halls stand close to the towers, open to light and sound. Here, children learn geometry through tone: a line sung becomes an angle drawn. Small harmonic boards scatter dust into shapes according to the notes they sing, revealing the visible structure of sound. Language and number follow naturally once the pattern is understood. Competition is unknown. Learning is collective, guided by rhythm and curiosity rather than reward.

Daily life follows the same calm order. Markets are held on the bridges between towers, where the air is cool and steady. Goods are moved in silent lifts that rise and fall through electromagnetic shafts. Exchange is conducted by resonance token, small discs carrying a recorded tone that identifies the maker and the measure of labour. Each token is re-tuned at the year's end to keep the economic field balanced.

Every seventh day, the whole city gathers for the Renewal: a low communal hum begun by the Circles and joined by every citizen able to voice a note. The combined vibration refreshes the local field and clears minor distortions from the week's activity. When the hum subsides, the city feels lighter, the air brighter. It is as ordinary and necessary as cleaning a hearth once was in later ages.

As night falls, the towers change colour. Their outer sheaths store energy through the day and release it as soft luminescence. Streets glow amber, domes blue-white, walkways silver-green, according to the minerals embedded within them. There are no lamps, no shadows cast by flame, only the calm light of resonance itself. The city sleeps, yet the grid never rests. It breathes steadily through the hours until the next dawn's calibration begins.

I watch them and remember that even perfection carries tension. The slightest drift in a tower's pitch is enough to register across the field, and already the Tuners sense faint irregularities in the deep tone of the Heart below. They adjust. They compensate.

The song holds, steady and responsive. The change is too small to name, yet I felt it through every layer of the realm. In that nearly inaudible tremor, the long story of the Fall has its first stirring.

To understand what was about to be lost, you must know the people themselves, the First Humanity, whose bodies and minds were shaped by the harmony they once held.

Chapter 11 - The Tartarians Themselves

They were the tall kind, the First Builders, stewards of balance. Broader of shoulder, longer of limb, and calmer in spirit than those who came after the Fall.

They were not the same species that walks the Earth today, but the First Humanity:
a high-coherence branch of the human soul-line, distinct in form yet ancestral in essence.

Theirs was a race apart in body, shaped by ages of living in accord with the Field. The Aether marked them deeply. Its ceaseless resonance refined bone, blood, and breath until endurance and clarity became their nature.

The Tartarian physique was remarkable for proportion. The average height stood tall, the torso long and upright, the skull broad and high at the crown. Hair tended toward pale tones that caught the light of the towers. Copper, flax, and silver-grey were common. The eyes showed faint mineral hues drawn from the region's soil. Green where malachite lay beneath the hills. Blue near lapis veins. Amber in the copper valleys.

Their clothing followed function rather than wealth. Outer garments were woven from resonant fibre, a blend of soft plant thread and trace metal spun to conduct the body's field. Men wore layered tunics and sleeveless coats fastened with flexible copper clasps. Women favoured longer robes belted with woven silver cord. Shoes and boots were supple, lined with conductive mesh so the wearer remained in tune with the ground. Every thread served a purpose, yet the overall effect was graceful and deliberate. Elegance born of precision rather than ornament.

Food was simple and pure. The soil itself was alive with charge, and crops grew dense and mineral-rich. Roots and grains formed the base of every meal. Fruit and leaf provided the rest. Small herds grazed freely on the terraces, their milk used for nourishment and medicine alike. No flesh was taken from animals whose frequencies supported the grid. Each species was believed to hold a place within the greater harmonic body of the Earth. Their diet kept them strong yet light, able to work long hours without fatigue or dullness.

Life among them was ordered by skill, not birth. Each person belonged to a Guild of practice. Builders. Gardeners. Healers. Artisans. Scholars. Movement between guilds was free. Aptitude decided the path. Harmony was the law.

Disputes were rare and settled by tone. Opposing sides would sound their argument before the council, each presenting a sequence of notes representing intent. When the frequencies aligned, agreement was reached. It was reason expressed as sound.

Leisure was not separate from labour. Music, craft, and study were woven into daily rhythm. Evenings often ended in shared song within the public halls, voices blending until the resonance matched the city's key. The Tartarians held that sound kept the mind clear of distortion. To sing together was to cleanse the communal field.

Instruments were simple. Stringed frames. Copper flutes. Hollow drums. Yet mastery of timing and tone could move the air for streets around.

Education extended through life. Children learned by observation, apprenticeship, and play within the tone gardens. Adults studied through the guild houses, refining craft and deepening their understanding of resonance. The old were revered not for authority, but for steadiness. Their presence stabilised the younger fields around them. Age touched them slowly. Many lived several centuries, their vitality maintained by alignment rather than remedy.

Belief was practical. They saw the Earth Heart not as deity, but as living mechanism. The pulse of existence to which all beings owed balance. Gratitude replaced worship. Their gatherings were acknowledgments of the Field's constancy. Offerings of voice and light rather than sacrifice. When storms passed or harvests ripened, the people faced the tower, breathing in unison for three cycles of the pulse, then returned to work. It was devotion without superstition.

As for their cities, not all that is called "Tartarian" belonged to them.

Much of the surviving architecture, with its wildly differing styles and scales, arose from multiple civilisations that fell together during the Great Collapse of the timeline. Tartarian structures remain the most refined, but they stand layered beside the ruins of other ages. Each swallowed by the same catastrophe. Each later misunderstood as one.

In all things, they valued quiet competence.

Pride was seen as noise. A distortion of tone.

The perfect citizen was one whose presence calmed the Field.

Such was the measure of virtue in Tartaria.

A civilisation that mistook balance for permanence.

Unaware that even the most perfect harmony must one day change.

(Note: The diversity of ruins seen today does not reflect Tartarian-era cultural variety. Many structures originate from other realms and layers that collapsed downward during the Fall, merging into the same archaeological stratum.)

Chapter 12 - The Harmonic Alphabet - The Language of Tone

All Tartarian knowledge was built upon vibration. Their science was sound, and their language was its geometry. Where modern tongues rely on letters and sound-shapes, the Tartarians wrote in resonant glyphs, sigils that were not merely read but heard by the mind. Each symbol held frequency. Each phrase created tone.

Every line of script shimmered faintly when spoken aloud.

The Harmonic Alphabet formed the foundation of their civilisation. It was not invented. It was remembered. A rediscovery of the frequencies by which the world itself had first been sung into being. The alphabet contained sixty primary tones, each with subharmonics that could merge or divide like notes of a grand chord. To write was to tune reality. The most skilled scribes, known as tone-smiths, could shape matter through script alone, drawing patterns that would hum themselves into form.

Buildings, songs, and even daily conversation carried this subtle order. The names of people and cities were chosen with care so that their spoken resonance aligned with the surrounding grid. When a newborn child was named, it was not fashion that decided but frequency. The child's breath and heartbeat were measured, then matched to a glyph sequence that would harmonise with their inner tone.

Within the Tonal Guilds, every craft was learned through this alphabet. Physicians used it to balance organs and energetic flow. Architects inscribed it into the foundations of stone.

Artisans of light embedded it within crystalline metals.

Even the dragons and the giants perceived its harmonics, though each species expressed it in its own octave.

When distortion entered the grid, the alphabet itself began to falter. Some glyphs no longer glowed. Others produced broken tones. The people continued to write, but their words carried unease. The final days of Tartaria were marked by silence in the libraries, not from neglect, but from fear that a single wrong tone might undo what remained of balance.

A few symbols survived in secret, carved deep into the resonance towers where no flood or flame could reach them.

They hum still beneath layers of soil and salt. The original language of creation, waiting for hearts once more tuned to hear.

Chapter 13 - Crystal Libraries and Living Records

Knowledge in Tartaria was not written. It was sung into stone. Every city held at its centre a crystalline hall whose walls shimmered faintly even in darkness. These were the Living Records, repositories where tone itself preserved history.

The archivists, called Resonants, used faceted rods of clear quartz to project controlled frequencies into great crystal pillars. Each vibration encoded layers of information, not in words but in harmonic patterns. When played back, the pillars re-emitted light and sound, recreating the original event in full sensory clarity.

To stand within a Record was to witness the past directly.

The memory lattice was powered by the same Aetheric flow that fed the grid. As long as the towers sang, the libraries remained alive, slowly learning from every tone that passed through the city. In this way, knowledge was not a possession but a living conversation between generations and the field itself.

The Resonants guarded their craft with devotion. They could summon a forgotten formula by hum alone, or call forth the image of an ancestor through the vibration of a single note. When the Empire fell silent, many of these libraries petrified, their surfaces clouding into opaque quartz. Some survive still as what later ages call singing stones, faintly responsive when touched with the correct pitch.

Even now, where ruins hum beneath moonlight, it is said the crystals still remember. Their tones are trapped like held breath, waiting for the world to resonate once more.

Chapter 14 - Clothing, Tools, and Everyday Craft

In Tartaria, even the smallest article of daily use served the field. What later ages would call fashion or ornament was, in truth, a continuation of engineering on a personal scale.

Clothing was designed to stabilise the wearer's resonance.

The base fabric came from a fibrous plant grown in charged soil, spun with fine threads of copper and silver. The metal strands formed an invisible mesh that conducted the body's natural current, keeping pulse and breath in phase with the city's tone. Garments were layered rather than thick; warmth came from the circulation of energy, not from trapped heat.

The cut followed geometry rather than trend. Tunics and robes were proportioned to the same harmonic ratios used in architecture: shoulder width to sleeve length, collar arc to hem radius. When correctly made, the wearer felt lighter and unstrained in movement. Colour was secondary to frequency. Each dye mineral carried a measurable wavelength. Builders wore earth hues to ground their field. Tuners and engineers preferred pale blues and greens for clarity. Gardeners favoured copper-red, matching the soil's charge. Nothing was symbolic. It was practical resonance management.

Footwear completed the circuit. Soles were layered with conductive mesh and fine crystal dust sealed between pliant leather sheets. Every step renewed contact with the ground flow.

Travellers carried small grounding rods that could be pressed into soil at rest stops, discharging static before sleep. Such habits prevented the fatigue that long journeys would otherwise bring.

Jewellery in this age was instrumentation. Rings acted as micro-tuners, containing adjustable inserts of quartz or tourmaline to correct personal frequency drift. Pendants served as stabilisers, worn close to the heart where the field was strongest.

None of it existed for vanity's sake. The clearer one's tone, the easier it was to work, think, and rest.

Tools followed the same logic. The average citizen carried a compact tuning rod no longer than the hand, used to check surfaces for harmonic stability. Builders used longer rods for walls and vaults. Artisans employed delicate probes that emitted a faint hum when alignment faltered. Every workshop held a resonance plate for recalibrating tools weekly. Even domestic utensils such as pitchers, bowls, and lamps were shaped to distribute vibration evenly so cracks never formed from standing waves.

Markets reflected this craft culture. Stalls offered replacement fibres, tuning crystals, and small repair kits. Families maintained their own garments and tools, patching and re-tuning as needed. Nothing was discarded while its pattern remained intact.

Age added value. A well-kept robe or instrument, harmonically seasoned by years of use, carried a steadier tone than any new piece.

Children learned these skills early. In open courtyards they practised threading copper through cloth, listening for the faint whisper that confirmed continuity. They learned to judge true resonance by ear, striking a bowl or beam and waiting for the clean decay. By maturity, every citizen could repair the tools of daily life and keep their own field balanced. Dependency was unknown. Maintenance was a form of citizenship.

To outsiders, these habits might have seemed obsessive.

Within the grid, they were natural. Stability of the whole began with the stability of each life. Thus, the humble act of sewing a seam or polishing a ring was not vanity, but participation in the living harmony of Tartaria.

Chapter 15 - Movement and Transport

To move within the realm was to join the current itself. Travel was not forced against resistance but guided by flow. In Tartaria, movement obeyed the same laws that governed light, tone, and breath.

The sky-trains were the arteries of the empire. They ran upon suspended magnetic rails that circled the high cities and spanned the great plains between them. Each line followed a path mapped to the ley geometry beneath, never straight, always curving with the natural rhythm of the grid. The carriages were formed of light alloys lined with crystalline conduits that drew energy directly from the towers. There was no smoke, no wheel, no friction. They hovered a hand's breadth above the rail, held steady by alternating fields. Once moving, they required almost no power to maintain speed.

At the terminus of every city stood a harmonic gate, a tall frame of copper and quartz that linked the line to the local resonance field. When a train arrived, the gate synchronised its frequency to the city's tone, equalising the pressure of the Aether before passengers stepped off. The exchange took seconds. Travellers spoke quietly while waiting, as the air around the gates vibrated softly, and conversation above a whisper could cause discomfort.

Routes varied in altitude according to terrain and node density. Over lowlands the tracks floated a few dozen spans above the fields. Across mountain regions they rose higher, following the sky bands where air resistance was least. Bridges and pylons were built of the same living geometry as the towers, their bases anchored to sub-nodes tended by giants in the tunnels below. Maintenance was continuous yet effortless. The system tuned itself, using feedback from the moving carriages to adjust field strength along the route.

Short-range travel within cities relied on moving pavements. These were narrow belts of magnetised stone running parallel to the streets, carrying pedestrians at a steady walking pace. To step on was to glide; to step off was to resume one's own rhythm without pause. The belts drew power from the local tower grid and never stopped. Elderly citizens and children used them freely, with no need for beasts or wheels.

Over open ground between the tower networks, smaller craft served as individual transport. Gliders rode the Aetheric updrafts, their wings of thin metal mesh vibrating in response to atmospheric tone. Field-sleds used short pulses of magnetism along the soil, leaving no tracks behind. Merchants and surveyors favoured them for moving between farming terraces and lesser nodes.

Goods moved with equal grace. Freight trains of larger size followed the same elevated routes as the passenger lines, their loads contained within resonance-sealed compartments. Perishable produce remained fresh for weeks, as vibration replaced refrigeration. Dock platforms at the city rims were arranged in vertical tiers: sky-trains on the upper level, gliders and sleds below, all timed by the same pulse transmitted from the central tower.

Travel was calm. There was no crowding and no haste. Departure times aligned with the planetary rhythm, and delay could not exist because all movement formed part of one continuous circuit. Journeys between cities that would later span continents took only hours, yet travellers seldom marked duration. They measured only the change in tone as one region's field merged into another.

Maintenance of this vast network required coordination across every guild. Engineers monitored current strength. Tuners ensured the harmonics of rails and pylons remained true. Sky-navigators adjusted schedules according to atmospheric flux. Failures were rare, and when they did occur the grid compensated automatically, rerouting current around damaged sections until repairs were complete.

At night the sky-trains glowed faintly, each carriage a line of light sliding through the darkness. From the ground they resembled slow-moving stars tracing deliberate paths across the horizon. Children watched them pass and knew, even without seeing, that every note of the realm's great song was in motion, and that the world itself was travelling with them.

Chapter 16 - Undersea Cities & Oceanic Towers

Beneath the mirror of the sea lay whole continents of stone and light. What the surface world would later mistake for coral ridges or drowned temples were once the living arteries of Tartaria's maritime grid, vast undersea cities built where resonance lines crossed the planetary waters.

The oceans, too, were tuned. Each current carried frequency in the same way air carries song. Along the floors of the Pacific and Atlantic stood radiant towers, crystalline pylons glowing with pale golden fire, harmonically linked to those upon the land.

Through them the Aether flowed without resistance, maintaining equilibrium between water and air, pressure and tone.

The coastal Tartarians were stewards of these depths. From ports that now lie silent, beneath what are now Japan, the Caribbean, and the Azores, they travelled downward through shafts of blue light in transparent craft formed of living glass.

Their vessels moved not by propeller or flame, but by harmonic counterweight, balancing sound and Aether to sink or rise through the sea as easily as breath moves through the lung.

In the Arctic Rim, where the grid's pulse was strongest, stood the Vault of Lyun, a circular city grown from pearl-like stone.

There the resonance of the polar currents kept the oceans clear and temperate, defying the cold of the outer world. Its towers, now half-buried in ice, still radiate a subtle hum. Divers and explorers have heard it, a deep tone that seems to call the heart home.

When the inversion struck, the water sang first. Whales panicked and currents convulsed. The deep resonance shattered into chaos, imploding whole chambers of glass and stone.

The light went out in stages, first blue, then green, then gone. Above, the surface cities felt the shock as earthquakes and rising seas, sensing that something vast had failed below them, though they could not yet grasp what was being lost.

Yet Astra remembers. She has seen the glow that still lingers, faint but unyielding, in trenches and caverns where the sea touches the buried grid. These remnants continue to hum, holding fragments of the world-song until the Flame returns.

And when the oceans again sing in tune, those lights may rise, not as ruins, but as lanterns for the next dawn. Above those sleeping seas, the towers once mirrored their deep counterparts in silent correspondence.

Chapter 17 – The Gate of Twelve and the Energy Nodes

The Gate of Twelve was not a structure in a single place, but a system that spanned the realm. Twelve principal Energy Nodes anchored the harmonic grid of Tartaria, vast confluences where the power of the Aether met the pulse of the Earth Heart. Their alignment formed a single living circuit, the twelve notes of the planetary chord.

Each node was crowned by a tower complex tuned to one key of the scale. Together they produced the great Resonant Crown, a field that wrapped the known world in equilibrium. When the towers sang in unison, the atmosphere brightened and the white girdle at the world's rim glowed with blue fire, proof that the current was whole.

The Gate itself existed at the harmonic centre where all twelve waves met, a focus known only to the highest tonal guilds.

Through this convergence, Tartarians could channel immense flows of energy, opening what they called the Breath of Creation, a field that joined higher resonance bands to the physical plane. In those moments of perfect alignment, the air shimmered like glass, and the boundaries between thought and matter softened until they dissolved.

Each node also governed a realm of purpose: growth, healing, memory, movement, balance, light, sound, form, time, water, fire, and spirit. The citizens nearest to a node often developed a natural aptitude for its domain. Healers gathered near the green node of growth, mathematicians and archivists near the golden node of memory. In this way, diversity arose without conflict, each region expressing a single aspect of one complete chord.

The Gate of Twelve has never closed. It merely sleeps. Beneath the soil, the nodes still hum, waiting for the world to find its tone again.

Chapter 18 - The Giants

They were not myths but neighbours, sentient guardians standing between twenty and thirty feet in height, their stature varying by lineage and region. Up close, their skin held fine fractal ridges like etched maps of the ley lines below. When a hand met that surface, it thrummed softly, like distant thunder.

Those who lived alongside them said that a giant standing still was the calmest thing in the world.

Work in the Living Grid

The giants' role was balance, not rule. They served as harmonic anchors attuned to towers and ground currents.

They stabilised the towers. Long before instruments registered any deviation, giants felt pitch drift in their bones. With a palm placed against a column and eyes closed, a structure would settle as though reassured.

They shielded during Aether storms. When charge tore loose from the sky, giants stood between the people and the surge, grounding energy through their own bodies. Silver lines etched across shoulder and arm recorded such nights, marks of service rather than injury.

They read the waters. At dawn, giants walked the resonance canals and could discern a current's cause by its motion and colour alone. Instruments followed their judgement, not the other way around.

They offered quiet companionship. A giant at rest could calm a troubled nervous system by presence alone, carrying a fragment of the world's original peace through the streets.

Names Carried Forward

A few are remembered with clarity. Calomir, the stone-voiced warden who steadied the Northern Spire for centuries. Mesa, river-walker and listener to underground streams. Hahrun, sentinel of the Eastern fields and friend to children.

Every towered city was guarded by one or more such beings.

To lose even one would have ruptured the culture. What came was the loss of all.

How They Were Regarded

No one called a giant "it." Gifts gathered along their routes: braided grasses, carved tokens, polished stones tucked into wall crevices where they passed. Elders sought their counsel before major decisions, not for politics but for the giants' instinctive understanding of land truth.

The Still Sentinel

In later chronicles, one name returns again and again: Solkaan, petrified above a fault rift, standing watch over the Resonance Cradle that shelters a dormant fragment of the Earth Heart. It is said that if stone should breathe once more, if even a hairline crack opened at his foot, the hidden chamber would awaken, and the Stone Sign, herald of renewal, would begin. Even in silence, he guards.

A Note on Measure and Presence

I saw them standing between twenty and thirty feet in height, though their measure varied by region, and some smaller kin worked nearer the surface. Yet what set them apart was never their size, but their stillness. Through that quiet strength they held the field in balance, allowing all life above to move in peace.

I mark them here because to understand the Fall, one must first know what was asked of those who tried to hold the frequency when the tone began to change.

As the grid's coherence softened, this communion became less exact. The shared field no longer aligned without effort, and meaning that once travelled cleanly through tone began to require greater care. Emotion still carried resonance, but it no longer translated with the same immediacy. The Choral Exchange endured, though increasingly as a discipline rather than an instinct.

After the Fall in secluded places, deep caves, mountain towers, and quiet sanctuaries, the practice was maintained in its older form. There, the universal language was kept alive, not in defiance of change, but in remembrance of what unity had once felt like.

Chapter 19 - Inter-Species Communication Systems

Unity in Tartaria was not imposed. It was felt. The many beings of the realm, Tartarian, Giant, and Dragon, spoke languages of their own, yet all could communicate through a single medium: resonance.

In a world where every living being carried its own harmonic signature, the art of communion lay in aligning one's tone to another's. Words were seldom required. A Tartarian healer could calm a frightened dragon by humming a minor third octave.

A child could draw gentle warmth from an animal simply by matching breath and rhythm.

This practice was known as the Choral Exchange, a discipline taught to every child once they could speak. It was both speech and song, both science and spirit. By modulating frequency, Tartarians could send meaning directly through the Aether, an instant translation free of ambiguity or deceit. Truth, in Tartaria, carried a tone. Falsehood carried none.

The giants communicated in subharmonics that travelled through the ground itself. When they spoke across distance, the soil trembled softly, and the towers registered the vibration. Dragons answered from the air with long, drawn chords that spiralled through cloud layers and carried for leagues. The resonance towers acted as receivers, catching these vast signals and rendering them into patterns the smaller races could perceive.

When distortion entered the grid, this communion faltered.

The shared field fractured, and meaning blurred. Where emotion had once travelled effortlessly through tone, misunderstanding crept in like static.

Now, as the Flame begins to stir once more, faint echoes of the Choral Exchange awaken in the hearts of those who remember.

The song of understanding was never lost. It has only waited to be heard again.

Chapter 20 - Sacred Geometry and Ley Design

To the Tartarians, geometry was not mathematics but language, the visible grammar of creation. Every curve, angle, and ratio echoed the pattern of the Aetheric field. By building in obedience to these laws, they invited the world's energy to remain stable and clear.

Their Ley Design formed the skeleton of the realm. Lines of telluric current, invisible yet measurable by tone, connected cities, towers, and waterways. Where three or more lines crossed, a node was born, and from each node a geometric figure unfolded: hexagons for balance, triangles for flow, circles for containment. The whole planet became a lattice of living mathematics.

Structures were laid out according to the Law of Twelve and One, with twelve radiating paths converging on a central axis.

From temple to workshop, proportion was sacred. A doorway might mirror the ratio of a mountain peak; a bridge's span could match the harmonic distance between two star clusters. This unity of form and frequency kept the grid coherent. It was said that when a building's geometry was true, it sang, inaudibly yet perceptible to those attuned.

Geometry also guided travel. The ley corridors acted as highways of resonance, allowing vehicles and sky-trains to move without friction. They did not push against the air but rode the standing waves of the earth's own field. These same corridors carried information, pulses of light travelling across continents faster than spoken speech.

In art and dress the pattern repeated. Spirals, nested triangles, and twelve-point stars adorned garments, pottery, and even the cadence of poetry. These were not symbols of belief but reminders of the grid's living design. The people understood themselves as notes within a larger mandala, each movement and breath part of a geometry far greater than any individual life.

Geometry was the first to register the change. Angles that had once aligned without effort began to require correction.

Towers lost none of their function, yet their hum grew less certain. Light no longer passed through crystal windows with perfect consistency. The pattern still held, but its ease was gone.

Chapter 21 - Healing Sciences and Tonal Medicine

In Tartaria, health was not fought for. It was maintained through tone. Every citizen learned from childhood the discipline of harmonic alignment, the art of keeping body, thought, and field in tune with the Earth Heart. Illness was not considered invasion, but interference, a discord within personal resonance.

The Healing Chambers were circular halls built of white calcite and crystal. Sound moved through their walls like water. The patient lay at the centre upon a lattice of gold and copper threads while the healers, known as Tonics, surrounded them in concentric rings. Using calibrated chimes and their own voices, the Tonics produced frequencies that matched the body's living pattern.

When tone and tissue aligned, pain eased. There was no incision and no drug, only resonance restored to order.

The chambers were sustained by the same Aetheric grid that illuminated the cities. Each hall was connected to a Harmonic Tower, which carried the healing tones outward, creating regions of calm balance around the cities. Travellers passing through these regions felt lighter and calmer, as if burdens had been gently tuned away.

The Hydronomists, who governed the water grids, also played a vital role. They understood that water could record vibration. Before entering the canals, water was sung to in triads of love, clarity, and renewal. When consumed, it carried those frequencies into every cell, refreshing the body from within.

The Tonal Guilds trained diagnosticians who could sense disharmony through touch or sound alone. They used luminous charts that displayed the body's chords, organs mapped not by anatomy but by frequency. A healer would hum a scale and listen for resistance, then retune the note until flow returned. To them, the heart was not a pump but a resonator. The bones were antennae. The skin was the border of a song.

Even death held no terror. When a life reached its natural resolution, family and kin gathered to perform the Quiet Tone, releasing the soul gently into the wider resonance of the Earth Heart. It was said that a spirit departing in harmony would guide new life back into the grid.

Astra speaks:

The art of healing was the art of being. Harmony was life itself, and when a tone wavered, the soul simply paused to remember its song.

Chapter 22 - Names that Walked the Realm

The Giants (stone, depth, and pulse)

The North-Vein Listeners

This was not a family name but a duty-line. These were the tall attendants stationed along the northern conduit galleries, those who could hear sub-node drift before instruments registered it. Their discipline was to sit with a palm to the pillar and mark any deviation in pulse against the city's hour-tone.

In the period before the collapse, it was a Listener who first recorded a persistent irregular knock deep beneath the tenth mile, a deviation too faint to trigger alarms, yet steady enough to warrant note. That record would later become one of the clearest early indicators that the deeper harmonics were no longer settling as they should.

Stone-Binder of the Copper Gate

A single giant remembered for precision repairs. He taught apprentices to "close a crack by closing a ratio," reseating crystal shims until the pillar's note returned to the fifth. During routine calibrations, his habit was to stand silent beneath the tower, hands open, listening for the return hum rather than watching the gauges.

When a cross-tone first appeared in the Copper Gate's resonance, he felt it before it reached audible range. His warning reached the upper terraces seconds before the plates over-sang, allowing stabilisers to brace the arrays before the surge travelled further.

The Three Span Walkers

A small team assigned to long corridor runs between auxiliary nodes. Their pace was measured, three spans per breath, to keep heart and field steady within the narrow galleries.

They are remembered because one of their route-maps survived: a slate bearing harmonic notations and pressure markings recorded during ordinary transit. In the years that followed, those markings proved critical in reconstructing how the network had begun to lose synchrony long before the collapse became visible.

The Dragons
Sky-Bands, Weather, and Balance

East-Band Wardens

These dragons held to the pale blue stream above the eastern terraces. They flew repeat paths to smooth shear between bands so that the suspended rails rode clean.

In the final steady season, their flight patterns shifted subtly. Altitude changes occurred more frequently than before, a detail logged by relay clerks during routine traffic coordination and preserved in municipal records without comment at the time.

The High-Arc Pair

Two elder fliers who worked in counter-rotation above the mountain lines where tracks climbed toward clearer air. Their discipline was to trade height on the hour-tone so winds braided rather than struck.

On one recorded cycle, they broke habit and rose together, leaving the ridge momentarily unguarded. Sky-train logs note a brief flutter in the lanes during that interval, an anomaly that resolved quickly and was filed as transient atmospheric variance.

Low-Mist Keeper

A smaller guardian assigned to the fog channels above the river flats, thinning vapour so field plates remained dry and stable.

After the Fall, surviving photographs of the district reveal a subtle shift in masonry hue along the Keeper's former route.

The change aligns precisely with its patrol path, offering one of the few surface-level indicators of how deeply environmental harmonics had once been managed.

Chapter 23 - Bridges of Air and Light

The towers made it possible for the Tartarians to draw the sky down to them. In time, they learned how to rise into it. The hum of the resonance grid was constant, a living breath beneath their feet, and the people began to follow that breath upward.

From the high cities, great rings of copper and crystal extended outward, shimmering with unseen power. When their frequencies met the harmonics of the upper air, the atmosphere itself became a road. The first Aether Craft were born, vessels shaped by sound, guided by thought, sustained by the invisible currents of the Aether.

They lifted in silence, not by force but by agreement with the air itself. Each ship was a chord made solid, its hull a living instrument tuned to the pulse of the Earth Heart. To pilot was to listen. To travel was to harmonise. The sky filled with drifting lights that mirrored the canals below, and the world seemed to breathe in rhythm.

Beneath the oceans the pattern repeated. Towers of pearl and glass glowed in the deep, their currents echoing the movements above. In those undersea sanctuaries, Tartarians learned that water, too, could carry tone. The same geometry that guided the Aether Craft through air guided crystal vessels through the sea. Land, sea, and sky formed one design, one song of motion.

The fleets connected the realm. Trade, pilgrimage, and knowledge moved through the great web of harmonics.

To travel from one continent to another was to follow the melody of the planet itself. Giants listened from the plains and dragons circled in answer, their wings striking chords that steadied the air. It was an age of perfect balance, Astra's most luminous memory of movement made grace.

Then, faintly at first, the tones began to waver. A softness entered the hum, as though the breath of the world were catching in its throat. The towers flickered. The currents trembled. Astra felt it even before the Tartarians did, a change in the harmony, a single note beginning to fall out of tune.

She would remember that tremor forever. It was the first sign that perfection was not eternal, that even the brightest chord can fade.

Chapter 24 - Aether Craft - The Sky Vessels of Tartaria

Above the vast cities of Tartaria, flight was no miracle. The sky itself was a field of energy, dense with Aetheric current. Where later engines would struggle against the air, Tartarian craft moved with it, gliding upon invisible harmonics tuned between the resonance towers.

These vessels, known as Aether Craft, ranged from small courier orbs to immense sky arks capable of carrying whole guilds between continents. They possessed no wings, no propellers, and no fuel. Their hulls were forged of crystalline alloys that held frequency like instruments, each etched with the harmonic alphabet. When activated, the ship itself became a living tone, its hum lifting it from the ground.

Pilots were known as Tone Navigators, trained to align breath and heart rhythm with the vessel's inner chord. Through resonance they steered, not by force of motion, but by intention. A shift in emotion could turn a ship. A sustained note could raise or lower its course. The fleet moved gracefully through the lower atmosphere, linking the great cities with the remote towers that crowned mountain and plain.

Some craft could pierce the veil between realms. The Aether Gates, positioned high above the polar vaults, shimmered with bands of light through which only the most advanced vessels could pass. These journeys were not measured in distance but in vibration, each realm accessed through precise alignment rather than travel. Those who returned carried altered perception, having encountered mirrored structures, higher resonance bands, and forms of light that could not be stabilised within physical containment.

During the Fall, the sky itself turned against them. When the inversion spread, the currents reversed and flight became perilous. Whole fleets hung suspended before plunging silently to the earth. The great Ark of Miren fell into the northern seas, scattering luminous shards that still wash ashore as faintly glowing fragments.

Yet legends persist of vessels that did not fall. Some rose beyond the storm, following the last stable tone into the upper Aether.

They vanished from sight but not from memory, becoming known as the Wanderers of the Void. It is said that when the Earth Heart sings once more, they will descend again, shining, resonant, and whole, bearing knowledge of worlds unseen.

Chapter 25 - Daily Life: Work, Ritual, and Play

Above all else, Tartaria was a civilization that lived in rhythm. Every day unfolded according to tone, light, and the measured breath of the Earth itself. Time was not ruled by clocks but by resonance. Each district followed the modulation of its nearest tower, whose tone deepened and lifted with the planet's pulse.

The dawn calibration began each morning when the first shaft of light touched the upper terraces. The city stirred at once. The Women's Tuning Circles sang their greeting tones to the towers, the sound rising through the streets like a gentle tide. Men and women paused wherever they stood, in craft halls, gardens, or markets, and allowed the harmonics to pass through them. It was said that this moment kept the field clear of distortion and the mind focused for the tasks of the day.

Work itself was not drudgery. Each person held a task aligned to their frequency aptitude: builders, gardeners, healers, artisans, scholars, engineers. These roles interlocked like the geometry of the cities themselves. Labour was considered an act of participation in the grid. To lay a brick in the proper ratio was to tune a note. To polish a metal plate until it gleamed was to extend light into the world. Payment came through resonance tokens, small charged discs marked with a unique tone rather than stamped value.

A good deed and a good sound were one and the same.

Guild halls formed the centres of civic life. Each guild maintained a resonance room for collective tuning and instruction.

Apprentices gathered there to learn not only craft but the philosophy of harmony: that no sound exists alone, and that strength is the meeting of frequencies in balance. Between shifts, the guilds exchanged members to keep understanding broad. Gardeners studied building tone, engineers learned the health songs of the healers. Knowledge flowed freely, because secrecy was considered a fracture in the field.

The midday pause brought quiet across every district. Meals were taken communally in open courts shaded by vine and stone. Food was simple but abundant: roots and grains grown in charged soil, water drawn from condensation towers, fruit that glowed faintly in sunlight. The act of eating was another calibration. Each ingredient carried its own vibration, chosen to restore the body's alignment after labour. Music accompanied every meal. Even the smallest child knew the melodies that soothed digestion and replenished energy. Conversation was measured, cheerful, and free of haste.

In the afternoons, civic duties alternated with recreation. Children learned geometry by tracing spirals of sand in the tone gardens. Adults practiced forms of motion that blended dance and exercise, keeping breath steady with the rhythm of the towers. Games of skill involved harmonic instruments. Contestants tuned small plates to produce precise interference patterns, competing for the clearest sound. The point was not victory, but harmony restored after friendly discord.

Evening drew the people to the public terraces. The towers' inner light deepened to gold, and the air thickened with resonance. Families gathered for the dusk calibration as the Women's Circles once again sent their tones through the upper air. Children fell silent at the first note. Animals lay down in calm. The great hum that followed was felt through every stone and bone, a sound that unified body, city, and earth.

Afterwards came the leisure of storytelling. Scholars recited accounts of earlier ages. Musicians performed the sequences that mapped the creation of the grid. There were no theatres in the later sense. Every courtyard, every open space was a stage. Those who preferred solitude walked the sky-bridges, watching the faint luminescence of the sky-trains trace across the horizon.

Ritual and play were indistinguishable. To dance was to maintain flexibility in the field. To sing was to cleanse it. Weddings were harmonisations of two personal frequencies, performed under the guidance of the local Tuners. Births were greeted with the low tones of the elders to stabilise the child's field. Deaths were met with silence, allowing the departing resonance to fade naturally into the greater pattern. Nothing was mourned, for nothing truly left the grid. It merely changed pitch.

There were no prisons, for imbalance was corrected through re-tuning. The small number who strayed from order were sent to the quiet halls beneath the towers, where the air vibrated with soothing frequency. After a time, they returned healed of discord. Punishment was seen as futile noise. Only harmony restored could serve the whole.

In every way, Tartarian life moved with purpose. Each day followed familiar patterns, yet no two expressions were identical. Variation arose naturally within order, and balance was maintained without effort or vigilance.

The rhythms of daily life flowed cleanly through work, rest, and exchange. Harmony was not something pursued or protected.

It was simply how the world functioned.

Chapter 26 - The Steady World

Astra: I did not observe these days from afar. I was present within them. What follows is my remembering of the time before the Fall.

The Golden Node of memory between ages breathed beneath the city.

Far below the bright terraces and the sky-rails, its resonance caverns spread in vast rings. Stone shaped by ancient hands. Crystal grown to carry tone. A network of light-veins pulsing with the measured rhythm of Tartaria's grid. The air down there was warm and charged, never still, always humming with the low song of the world.

However, the grid was never perfect.

Every day brought its minor distortions. A slight drag along an outer line. A pulse arriving slightly distorted. A memory-plate projecting a blur that needed to be coaxed back into clarity. It was steady work, intricate work, endless work. For those who served the Node, this was normal.

Eryndor stood at the heart of it, a senior keeper of memory.

He worked on a raised platform surrounded by floating crystalline plates, each suspended on its own resonance. Geometry and light moved across their surfaces, mapping the state of the grid in shifting patterns. A soft golden glow ran along his forearms as his hands passed over them, guiding tone, smoothing fluctuations.

He was taller than most Tartarians, long-limbed, balanced, built for endurance rather than display. Under the amber cavern light, his skin held a subtle warmth, as if the Golden Node had soaked into him over the years. When he concentrated, faint arcs of resonance traced across his shoulders and sternum, a visible echo of the field he carried.

His face was calm and sharply defined. A strong jaw. High cheekbones. Eyes that held an amber-gold undertone when the light caught them. Not fierce. Not soft. Simply present. He had the look of someone who belonged where the work was hardest, not because he sought struggle, but because his presence steadied what surrounded him.

He wore simple tuning attire, layered harmonic fabric woven with resonance threads, pale and functional, trimmed with discreet bands that brightened when he engaged fully with a plate. Nothing was ornamental; everything was purposeful.

Most workers of the Node relied on instruments and training to read the grid.

Eryndor clearly felt it.

He could sense distortion without looking at the plates, by the way a tone sat wrong in his chest, by the pressure behind his eyes when a line drifted, by the way the floor seemed to tilt by a fraction when a distant node slipped out of alignment. It was not dramatic. It was simply how his field worked.

He had not been taught that.

I remember the first time I saw it.

It was years earlier, when he was still in training. I had descended in form to assist in the correction of a drifting line in the Seventh Node, my path took me through the Memory Halls and into a vast circular chamber called the Hall of First Resonance.

Amber-lit stone lined the walls. Crystalline plates were arranged in rings around a central core. Keepers learned to listen there, not with ears, but with field sensitivity.

He stood near the outer ring, hands hovering above a plate that emitted a low, uneven pulse. He was trying to correct a mild harmonic drift in the plate's memory projection.

His brow was furrowed, not in frustration, but in concentration. The teacher overseeing the hall was occupied with another student, so he worked alone.

I watched while he steadied his breath, placed both hands gently on the plate's surface, and spoke a single word.

"Settle."

The plate obeyed.

The resonance stabilised into a clean, steady tone. Too clean for someone who had barely begun formal training.

I paused at the entrance walkway, because the tone he drew from that plate was not a tone a beginner should have been able to hold. It carried coherence, not borrowed, not forced, but innate.

I stepped into the hall.

I wore the white-gold harmonic attire of the upper-band attendants, and my field was calm, aligned, resonant with Aether light, the air shimmering faintly around me because I had only just crossed from higher density into physical form. He turned slightly as the plate brightened, and that was the first moment he saw me.

He straightened himself. Hands moved behind his back, posture upright, like someone accustomed to respect.

I approached him slowly, deliberately, so my field could align with the density of the hall, and for a short time examined the plate he had stabilised.

"Who taught you that correction?" I asked.

His answer was immediate.

"No one," he said. "It felt wrong, so I adjusted it."

That was the moment I recognised what he was.

Not a promising student only, but a field carrying continuity resonance. A Keeper born to feel distortion as memory, not calculate it as error.

"What is your name?" I asked.

"Eryndor," he replied.

I watched him for a breath longer, then said what was true.

"Your field is unusual. Walk with me."

He hesitated for half a heartbeat, not from fear, but from assessing me. Then he followed.

As we walked toward the central core, I adjusted my resonance to match the density of the hall.

When our fields aligned, his steps faltered.

Not visibly.

Not dramatically.

Just a momentary shift, as though two puzzle pieces had clicked together.

He glanced at me with sudden realisation.

Not conscious memory.

Not knowledge.

Recognition.

He didn't understand why.

Not then.

Not fully.

But later, after years of working together in the deep chambers, he told me:

"When I met you, it felt like coming home to a place I didn't know I had forgotten."

That is exactly how it felt.

For both of us.

That was how and where I first met Eryndor. Neither destined nor dramatic. Simple and perfect.

Some time later, after training and alignment exercises, after his field had stabilised into its natural steady tone, he was assigned deeper. Not as a test, but as recognition. The Resonant Deep did not accept those who pushed or performed. It accepted those who could hold.

That day remains vivid to me.

The chamber beneath the Golden Node was dim, lit not by sun or flame but by amber luminescence emitted from the walls themselves. As he descended the spiral ramp, each footstep sent a ripple through the crystalline floor. His field responded instinctively.

The Deep stabilised him, and he stabilised it.

I was already there when he arrived, standing at the Source-thread interface, a vertical channel of white-gold light where the deeper resonance breathed into the world.

He stepped into the chamber, looked around once, and spoke quietly, almost to himself.

"This place feels like thought."

I looked up from the Source-thread and answered.

"That is because thought and memory are the same thing here."

He did not speak for a moment. He breathed the deep resonant air and nodded. Then he said, direct, focused, ready.

"Show me what you're doing."

He spoke with quiet certainty, fully present. I admired him instantly for that.

I brought him to the Memory Array, a long curved structure of crystalline plates arranged in a rising arc, each one glowing faintly with stored harmonic patterns. His task was to stabilise a minor distortion in the Memory Layer, caused by a drift that had brushed the Seventh Node.

The distortion was subtle, a faint stutter in the flow of old resonance. I placed my hand on the upper plate.
He placed his on the lower.

The moment our fields touched the same structure, the Array brightened.

His eyes widened, not from surprise, but from recognition.

He felt what few could feel. Not just the plate, but the relationship between our tones.

"Our fields complement," he said.

"They were designed to," I replied.

That was not metaphor. It was truth. Memory-Keepers of his lineage and Source Attendants of mine had been paired for thousands of years. The work depended on mutual resonance.

We worked in silence for several minutes, synchronising patterns, stripping distortion, guiding coherence back into place.

Then he leaned closer, studying a faint flicker running beneath the surface, and said with quiet certainty,

"This memory-band is older than the others."

I nodded.

"Yes. It predates the First Expansion. Be gentle."

His expression softened immediately.

He adjusted his field with such precision that the Array responded as if recognising an old friend.

When the Array brightened fully, he stepped back and said, calm as if reporting something obvious,

"It feels balanced now."

And I answered him, equally calm.

"Because you are balanced now."

He looked at me then with the first true recognition of a deeper connection, the one he would not fully remember until after the Fall, but which his field already knew.

That was his first true day beside me.

Calm. Precise. Harmonic. Unforgettable.

And now, many cycles later, he no longer stood as a trainee in the Hall of First Resonance. He stood in the Golden Node itself, one of its senior Keepers, his field fully matured and threaded into the heart of the grid.

I crossed from the corridor into the open space around the platform, my form fully realised, density matched to the Node's atmosphere, my field calm and aligned.

Workers in the nearby alcoves glanced up briefly, acknowledging me without surprise. In Tartaria, the presence of upper-band beings was not myth. It was rare, but real.

Eryndor looked up.

"Astra," he said.

I inclined my head.

"Namaste, Eryndor."

"Namaste," he replied.

We did not need ceremony. We had worked together long enough for the forms to be quiet.

"You've been here since first tone," I observed.

He nodded once. "The outer lines were becoming distorted today. Nothing unusual. Just more to watch."

"Distortion is increasing," I said, not alarmed. Only fact.

"It has been for some time," he agreed. "But still within what we can correct."

For a moment we stood in silence, the hum of the Node breathing around us.

The Golden Node continued to breathe beneath the city. On the surface, nothing had broken. Nothing had changed.

The first day passed in quiet steadiness.

Chapter 27 - The First Awareness

It was after a long shift in the Deep Chambers.

Eryndor was tired but focused, his field humming with the after-tone of grid work.

I invited him:

"Come with me.
There's something
you should see."

He didn't hesitate.
He simply said:

"Lead."

We ascended the high internal passages of the Golden Node, moving upward through layers of stone and living light.

The air changed as we climbed:

lighter,

cleaner,

more ordered.

His field expanded naturally as we neared the upper atmosphere.

We ascended through the inner passages of the Golden Node.

The corridors curved upward in slow spirals, stone giving way to lighter structures, amber tones thinning into pale gold.

With each level, the air shifted. It grew cleaner. More ordered. Less weighted by the deep resonance of memory and stone.

Neither of us spoke as we climbed.

There was no need.

When we passed through the final internal ring, the space opened suddenly, and the world revealed itself.

No human sky in later ages would resemble what lay overhead then.

The upper bands shimmered in layered order. Pale blue so soft it almost dissolved into silver. Gold drifting between currents like living mist. Faint rose spirals weaving through the higher air. The colours did not move at random. Each band flowed with intention. Each carried its own tone, its own rhythm in the greater breath of the world.

Dragons traced slow arcs through the currents, their long bodies cutting clean lines through the coloured air. Their wings stirred the bands without breaking them, leaving ripples that glowed briefly before settling back into harmony.

On the horizon, distant yet unmistakable, the Gate Stars glowed. Vast presences beyond the world, their light softened by the firmament into gentle luminescence.

Eryndor stepped forward to the railing.

He rested his hands on the stone and leaned slightly into the view. For a long moment he said nothing. His breathing slowed, deepened, as though something in him had remembered how to rest.

"I forgot how beautiful it is up here," he said quietly.

I moved to stand beside him, following his gaze across the layered sky.

"You never truly forget beauty," I replied. "You only forget that you knew it."

He nodded. I saw tension leave his shoulders, the kind that accumulates unnoticed over years of responsibility. We stood together in silence, listening to the soft hum of the upper air. The sound here was different from the Node below. Higher. Clearer.

A tone that loosened the bones.

Eryndor closed his eyes for several breaths and let the sky move through him.

His field brightened in response. Subtle golden arcs appeared at his shoulders, faint but unmistakable, as if the sky itself recognised him. He belonged here in a way few did.

"Is this what the world is supposed to feel like?" he asked.

"Yes," I answered. "This is how the Earth breathes when she is not wounded."

He opened his eyes and looked up again, taking in the ease of the bands, the way the dragons flowed with them instead of against them, the quiet presence of the Gate Stars beyond.

"I want to remember this for as long as I can," he said.

"You will," I told him.

And it was true. That memory would stay with him, even when much else was stripped away. It would surface in moments of quiet, as a sense of wrongness when the world no longer breathed as it once had.

We remained there for a long while.

Below us, the city moved through its patterns. Sky-trains glided along their currents. Towers exchanged tone. Tartarians rose into their tasks. Up here, the sound was softer, folded into the broader breathing of the realm.

Eventually, the light shifted.

He drew a slow breath and straightened, his hands still resting on the stone.

"If the sky ever breaks," he said, eyes still on the bands above, "promise me we will restore it."

I looked at him fully then. Not as a Keeper. Not as a worker of the grid. But as the being he was, carrying continuity deeper than he knew.

"We will," I said. "Together."

The words settled between us with quiet weight. A vow formed without ceremony, without witnesses, without the need to be spoken again.

We turned back toward the passage and began our descent.

As we moved downward, the air grew denser once more, the warmth of the deep chambers returning. Eryndor's field folded back into its grounded shape, but the imprint of the sky did not fade. It layered itself into him, a reference point he would carry without knowing how precious it would become.

Before we reached the lower levels, we diverted briefly to the upper observatory.

The chamber was circular, carved into the highest stone of the Node, its open aperture facing the sky. Light filtered through in shifting tones from the Life Band above, washing the walls in faint green luminescence. Crystalline columns displayed atmospheric patterns, pulsing slowly, rhythmically.

He tilted his head slightly.

"The balance feels off," he said. "Do you sense it?"

I moved toward him and placed my hand on the crystalline interfaces near the aperture, lifting my gaze toward the upper atmosphere.

"Yes," I said. "The Seventh Node is drifting."

He frowned, not in alarm, but in concentration.

"That should not happen this cycle," he said.

He was right. Drift of that kind belonged to transitions between great cycles, not ordinary days.

We moved to the stone seat near the wide opening where the sky beyond glowed faintly green with the Life Band and sat.

He rested his hands on his knees, his gaze distant. For several breaths, neither of us spoke. Green currents shifted gently beyond the aperture.

Then he asked, almost contemplative, a question he would not usually voice aloud.

"Astra… how long do you think we have until the next Great Shift?"

It was not fear that shaped the question. It was awareness.

I paused and answered honestly.

"Not long," I said. "But not yet."

He nodded slowly, accepting the answer with the quiet understanding of someone who had always felt cycles in his bones.

He kept his eyes on the sky.

"When the next shift comes," he asked, "will I remember this place?"

I turned my head fully toward him.

"Yes," I said. "You are built for remembering."

A small smile crossed his face, one of the rare ones, and said:

"Good.

I don't want to lose this."

The simplicity of that moment stayed with me.

He stood up to return to his work and paused in the doorway.

"Astra,"

"If things ever go wrong," he said, calm and certain, "promise me one thing."

"What?" I asked.

"Find me again."

My reply came without hesitation.

"Always."

There was nothing dramatic in the way he walked back toward the corridors. No tremor in the air. No warning tone sounded.

Just Eryndor returning to his work, the Node humming below, the world continuing as it always had.

The second day ended without rupture, but not without warning.

Chapter 28 – The Choral Ring

The next cycle did not belong to ordinary days.

The Golden Node still breathed, but its breath had shortened.

Eryndor felt it before the plates displayed it.

A deep, low tone rolled through the stone of the hall.

We knew that tone.

It was the unmistakable sound of the HAAR chambers activating emergency resonance.

He turned, not fast, but sharply, and moved toward the central stairwell. I followed a short distance behind.

We ascended two levels to the observation tier.

The giants were already there.

Three of them, standing in complete stillness around the main pillar. Vast forms of stone and living resonance, their presence usually a comfort, now an unspoken warning. Even they did not touch the structure. They watched it, grounded and braced, as the pillar pulsed with an unstable rhythm.

The air vibrated.

The walls seemed to breathe.

The central pillar brightened beyond its normal range, pale gold edging toward white. Not strain from load, but strain from imbalance.

Eryndor stepped to the railing and looked out.

The world beyond the aperture was wrong.

The sky bands, which should have layered cleanly in colour and tone, appeared dulled, flattened, as though the upper currents had twisted slightly out of phase. Not broken. Distorted.

He placed his hand on the guard crystal.

A shock ran through his arm.

He saw in a single flash:

the Form Node drifting

the Balance Node destabilising

the Water Node collapsing

the Firmament membrane twisting

the Thirteenth Gate beginning to stir

He whispered, not in fear, but in understanding:

"It is the Thirteen…"

The words settled heavily.

This was not local failure.

This was systemic.

As I approached the main chamber I could feel an uneasy pressure.

"Eryndor," I said, my tone steady. "Step back from the pillar."

He obeyed without hesitation.

Not because of authority.

Because of trust.

As he stepped back, the light surged. Pressure flooded the chamber, so intense that even one of the giants fell to one knee.

The Golden Node groaned, not in collapse, but in protest.

When the surge eased, Eryndor turned to face me.

"You feel it too," he said.

"Yes," I answered. "The drift has passed the threshold where local correction is sufficient."

For a moment, neither of us spoke. The hum of the Node filled the space between words.

Then he said softly,

"You need to go to the Choral Ring."

The words landed like a weight placed carefully on a table.

My field tightened in response, not fear, but in realisation.

I met his gaze. "That is not my place."

"It may be the only place remaining," he said quietly.

"If it worsens."

"If the deeper harmonics fail, holding here will not be enough."

"And you?" I asked.

His expression stilled. "I will remain," he said. "I will hold what I can."

"And if it worsens?"

He looked away. "Then you will be where the world can be best held."

I inclined my head, accepting what neither of us needed to say.

"Then I will go," I said.

He nodded. "You must"

The tension in his shoulders did not ease.

We stepped closer, near enough that our fields brushed briefly, a passing of resonance that carried more than language ever could.

"Hold," I said.

"Always," he replied.

I turned from the warm hum of the Golden Node and moved upward through the ascending passages. The stone light thinned. The air cooled. The sound of the world below softened into distance.

Some passages feel like stairs.

This one felt like stepping over a threshold that would not return.

Behind me, the Golden Node continued to breathe, steady still, but no longer unaware.

The world had not yet fallen.

But it had begun to lean.

Chapter 29 - The Great Fall

Astra speaks softly:
I have not spoken of this moment until now.
Some memories do not diminish. They wait.
Not because they are hidden, but because they resist words.
I will recall it to you as I experienced it, without distance or softening, as it truly was.

From the Choral Ring in the realm above, the ground beneath Tartaria trembled in a way no living layer should.

The Golden Node's song, usually ordered and steady, thickened under a weight it could not bear. Tones that should have flowed cleanly began to drag, as though the world itself had been forced to breathe against pressure.

I stood between Aurelith to my left, male in essence, steady and grounded, shaped in quiet gold. His field was dense with resolve, an anchor against unraveling forces.

To my right was Seravei, female in essence, finer and more fluid, her tone soft as luminous rain. She felt before she acted, her awareness moving like light through water.

Together, we formed the triad around the Choral Ring.

The great table of light.

The living map of the Earth-realm below.

It was no longer smooth.

No longer serene.

The layers flickered and tore, jittering like cloth stretched past its limit. Geometry failed to remember itself. Patterns rose and collapsed in rapid succession as harmonics slipped out of coherence faster than correction could contain them.

Working in silence, eyes closed, we wove stabilising tone into the fractures beneath us. Threads of resonance. Corrective forms. Entire architectures of balance summoned and released in moments. Each attempt encountered resistance without easing.

The collapse was already beyond reversal.

Then briefly, an essence touched my awareness.

It was Eryndor.

Unmistakable.

He felt impossibly close.

The recognition shattered a restraint I had held since the first fractures began.

I opened my eyes and stepped back.

The table before me still shone, its light no longer holding my focus. I lifted my gaze to the vast inner curve of the firmament that ringed the chamber, and felt my own resonance tighten.

Aurelith looked up at once, sensing the break in my focus. The triad was broken now, though it scarcely mattered.

The collapse below had already outrun us.

I crossed the expanse swiftly. Pausing as fear flickered, brief and sharp, at what I might feel of Eryndor through the seal. I placed my hand against the firmament.

As I did I felt his endurance, the measure of what had already been stripped away.

The barrier between the upper bands and the collapsing Earth realm was cold and unyielding.

I pushed resonance into it, searching for softness, for any thinning I could exploit. The surface yielded, then recoiled, hardening as quickly as I softened it. The seal was learning faster than I could adapt.

For a moment, I lowered my gaze. Then I looked back toward the table of light.

Aurelith met my eyes. He said nothing. He gave only the smallest shake of his head. Grief and inevitability held together in a single motion.

I turned to Seravei.

Feeling what I felt, living it within her own field. For a moment, we existed in the same pain. Then she inclined her head, once, slow and certain, and stepped forward. She placed her hand on the seal beside mine.

Together, we forced softening resonance into the barrier.

The firmament thinned under our combined tone, though it still crept back toward rigidity.

Aurelith stepped forward without hesitation. Placing his hand beside ours.

Our three fields aligned, forming a triadic harmonic. A tone that could not exist for any of us alone.

Our resonance braided together. Not as force, but as coherence.

A single place in the seal pulsed. It thinned. It softened.

A passage.

Brief & Fragile.

I looked toward Seravei. She returned my glance with a faint smile, compassion steadying her field. I turned to Aurelith. Sorrow still lay in his eyes.

I gave them both a single, wordless nod of gratitude.

Anchoring my field to the realm above, I did not hesitate further.

I stepped forward.

Light collapsed inward around me as I passed through the thinning membrane, my descent marking the realm below like a falling star. I struck the earth hard just beyond the Chamber.

The ground beneath my feet sagged and shuddered, as though it barely remembered how to hold form.

The world was not the one I had left.

Distortion was evident everywhere.

Structures leaned into themselves. The air trembled with unresolved pressure. The land folded and unfolded in slow, disorienting waves.

Raw distortion tore at my form. It cut at the edges of my field like blades of broken light. The ground shook with each movement.

I continued moving between falling stone and surges of fractured tone until I reached the Chamber.

Marukan, one of the larger giants stood at the pillar outside, his vast form braced against forces no physical structure should have been able to contain. He felt my approach at once not as presence, but as strain moving through the field he was already grounding.

He turned, eyes widening and shook his head in a slow, deliberate refusal. He understood immediately what it meant for those from the higher realms to be moving through the collapsing layers. He knew the potential.

I did not answer him.

There was no time, and no argument to be made.

Seeing my resolve he turned back to the pillar and widened his stance, drawing distortion into himself, forcing coherence where none wished to remain. His field became an anchor not a shield, but a conduit absorbing the impossible load so that the remaining structure could agree, for a few critical moments, to hold.

His sacrifice opened the path.

I crossed into the Chamber.

The walls pulsed with distortion, their surfaces rippling as though uncertain whether to remain stone or dissolve into something else. At the centre stood Eryndor, braced wide at the resonance table, both hands pressed flat against its trembling surface.

Shattered geometry flared upward in jagged bursts. White arcs of destabilised light tore from his shoulders and joints. His own field was breaking under the strain.

Before I could speak, the chamber convulsed.

A surge of distortion ripped through the walls and struck him, tearing his footing away. I moved without thought, my hand locking around his arm to brace his fall.

He turned, and our eyes met.

His gaze was deep beyond measure.

"Astra," he said "we are losing the grid."

Another wave gathered. I felt it building behind the chamber walls like a silent scream.

If I had not intervened his field would have been torn apart.

I placed my hand over the centre of his chest, anchoring my field into his.

When the distortion wave struck, it tore through him with merciless precision. I felt memory layers shear. I could feel his resonance shred beneath the force. I channelled everything I could, diverting the surge into the chamber's discharge zone below, carving out the narrowest margin in which he might endure.

The wave passed.

He still stood but barely now.

Still conscious & aware.

I leaned closer and spoke.

"**Eryndor, hear me.** I cannot stop the collapse. I cannot prevent your fall."

"**But I can choose how you fall.**"

He smiled softly, unafraid.

"**Then, Astra,**" he said. "**choose wisely.**"

Beyond the chamber walls, Tartaria was already answering the collapse.

The great bodies of the giants faltered as the harmonics beneath them failed. Pillars that had stood since the first shaping groaned and leaned. Those who carried the load of the world felt the ground betray them, not through weakness, but through forgetting.

In the skies, the dragons cried out.

Their resonance, once bound cleanly to the upper currents, fractured into dissonant spirals. Some fell in silence. Others burned as they passed, their descent marking the air with long scars of light and sound. The sky itself seemed to recoil from their passing.

Within the chamber, Eryndor's field wavered again.

His field would not survive the Fall as it was. Not as Eryndor. Not as he knew himself. The dimming of the Gate Stars had already closed the path of natural ascent.

If left whole, the inversion would strip him of every memory that made him who he was, tearing continuity apart until nothing coherent remained.

To preserve him, I had to fold him inward.

Not to erase, but to choose what would be kept.

To gather what could endure. To quiet what could not. To compress his field into a form that could pass through the lower bands intact, carrying strength without fracture, memory without collapse.

Grief rose in me like a tone I could not silence. A reluctant farewell to all he had been.

I relinquished my form, allowing it to give way to light.

It collapsed inward until only my field remained, a radiant core shaped by intent, wrapped in ordered layers that flared as they met the surrounding distortion. My field reduced to a coherent core, maintained against the surrounding collapse by alignment alone.

I wrapped my resonance fully around his being.

I began the compression.

Fractals of my field unfurled, weaving around him, accelerating, tightening. I could hardly bare what I had to do.

As I folded him inward, his physical body failed gently. It fell to the chamber floor as his field separated and rose, now mirroring my own. Wrapped in light. Held in motion.

I continued to compress him carefully. Care & sorrow threaded every movement. I chose what he would carry forward. What memory would survive. What awareness would sleep.

It was an act that could not be undone.

A choice made from care, and paid for in absence.

When complete, his field no longer resisted. It rested within itself, condensed to a single, shining seed.

I released him, gently.

He passed from my reach and entered the transitional layers, falling into the lower incarnation bands beyond my grasp. For a few silent seconds, the roar of distortion receded within my perception. I watched him drift away, carried by currents I could not follow.

The chamber screamed as the harmonics collapsed inward. Distortion surged back all at once, violent and unforgiving.

I was torn upward.

My field snapped back to its anchor in the higher realm, ripped through the closing layers of the sealing barrier as the firmament and the passage hardened now fully, cold and absolute.

I spoke his name. "Eryndor."

His name sinking into the barrier, unheard by the world below.

Behind me, the Choral Ring stood silent. Aurelith and Seravei did not approach. There was nothing to be said.

Below us, Tartaria continued its descent.

The giants fell where the ground forgot how to hold them. Dragons vanished into fire and shadow. Towers that had sung since the first shaping collapsed inward, their tones dissolving into noise and then into silence.

The Fall did not end in fire.

It ended in quiet.

In forgetting.

In long silence.

And in that silence, Eryndor began his passage into the collapsed world below.

Chapter 30 - The Convergence of the Layers

Astra reflects:

What is called the Fall was not destruction in the way later ages imagine it. There was no single explosion, no final moment that ended one world and began another. What occurred was convergence.

Before the Fall, the realms of Earth did not exist in a single place. They were separated by frequency, density, and harmonic order. Civilisations occupied different layers of the planetary field, each real, each physical to its inhabitants, yet largely unseen by those who lived outside its band.

Tartaria occupied the highest coherent harmonic layer of the world. Beneath it, other layers persisted, some older in formation, others simpler in structure, many already operating at reduced complexity long before Tartaria reached its height.
The grid maintained separation between these layers, preserving their distinct resonant states.

When the grid failed, the separations did not dissolve evenly. They collapsed.

The layers merged inward toward the lowest stable density. Frequency differences could no longer be maintained, and so reality resolved the only way it could, by compression.

This was the plunge into darkness.

Light did not vanish. It became inaccessible. The harmonic structures that allowed light to move freely between layers were gone. The world did not become night. It became unlit.

As the layers converged, entire cities arrived intact into the third dimensional plane. Towers did not crumble. Walls did not shatter. Architecture survived because it was coherent. What failed was context.

The interiors of buildings filled with matter that did not belong to them. Soil, clay, sediment, and crushed strata from lower layers were drawn inward as frequencies equalised. Lower levels filled first. Basements vanished. Stairwells became sealed chambers. Upper floors became ground floors.

Tartarian structures survived this transformation because they were harmonically proportioned. Their forms redistributed pressure through resonance rather than resisting it, allowing stone and vault to settle intact while surrounding material accumulated.

This is why later ages find advanced structures buried from the inside.

The ground level of the new world was not the ground of any previous one. It was an artificial plane formed by overlapping strata settling into a single compromise surface.

Memory collapsed faster than stone.

Consciousness does not survive compression intact. Beings whose awareness was adapted to higher harmonics were forced into lower density embodiment. The shock stripped memory, identity, and continuity. Not gradually, but immediately.

This is why the survivors did not remember what had happened.

It was not trauma alone. It was incompatibility.

Those who had lived entirely within the third dimension retained bodily continuity, but their minds were overwritten by the convergence. Their history vanished. Their sense of before was erased. They awakened into a world already old, already vast, already incomprehensible.

They did not question it. There was nothing within them that could. Deep amnesia and disorientation had stripped them of comparison and doubt.

Above them, the sky changed.

The layered firmament collapsed into a single atmospheric shell. The coloured bands vanished. The GateStar paths closed. What remained was a muted sky, dimmed not by clouds but by loss of access.

Below them, the deep places opened.

Fractures spread through the crust as harmonic pressure redistributed. Caverns that had been sealed for ages became reachable. Undercities emerged into accessibility. The grid that had once regulated passage between above and below no longer functioned.

This is the opening that followed the Fall.

The world after convergence was silent, dark, and malleable.

Not ruined, but undefended.

Civilisation did not end. It was replaced.

Those who moved into the vacuum did not need to conquer. They only needed to explain. To organise. To assign meaning to what no one remembered building.

The survivors accepted every story because they had no memory to contradict it.

This is the true legacy of the Fall.

Not destruction, but disorientation.

Not extinction, but erasure.

The world you inhabit now is not built upon ruins.

It is built upon a convergence.

And everything that followed, the orphaning, the infiltration, the rewriting of history, became possible only because of this moment when the layers became one, and the past became unreachable.

Chapter 31 – The Mud Flood and Burial

The empire did not end in a single instant.

The light faded, and then the earth began to breathe in another way.

When the towers fell silent, the ground beneath them loosened. The harmonics that had bound soil and stone together for centuries released their hold. The resonance that once maintained form now withdrew, and matter began to move as if it were fluid. Streets rolled. Foundations softened. Whole districts slipped downward while others rose and folded like slow waves.

In the north, the great city of Orath descended by three levels in a single shift. Market roofs vanished beneath fine silt. Upper balconies became doorways opening onto newly formed hills.

The shining halls of Vorn were buried to their lintels, leaving only upper windows exposed above the altered plain. Those who survived dug into the slopes the following morning and found their own homes sealed beneath them, intact and unreachable.

The giants, already stone, became markers of the new world. The tallest, Drenn Orisos, stood exposed only to the chest, his hands still resting on the half buried conduit he had been stabilising. To the east, Kaelor Jinn lay across what had once been the river channel, now reduced to a narrow flow threading between crystal ribs. Farther south, the smaller giant Aeron was completely entombed. Only part of a shoulder and a single hand remained visible until later ages mistook the shape for natural stone.

The flow carried away smaller life entirely. Herds were sealed within rolling banks of clay that hardened within days. Dragons, already petrified, became features of terrain. Wings formed ridgelines. Heads became cliffs. Their mineral dense bones channelled the last remnants of the grid into faint luminous veins that still glimmer beneath certain mountains when the sun strikes them at a low angle.

In the time that followed, survivors moved across the altered surface in silence and shock. The sudden compression of their fields had erased memory entirely. Identity, role, and history collapsed inward at the moment of impact, leaving those who lived in a state of profound amnesia.

Towers protruded at unnatural angles, their buried levels holding trapped air that moaned through fractures when wind passed over them. The Tuners spoke little, not from restraint, but because the knowledge that had once defined them no longer existed within reach. With the grid buried, their craft had vanished from thought itself.

Those who remained gathered instinctively, not by remembrance or duty, but by proximity and need. Children were kept close because hands reached for them, not because anyone remembered what a child represented. Language thinned, reduced to simple sounds and gestures. Whatever Tartaria had been, no one present could name it, or even sense that a name was missing.

The resonance had not been destroyed. It had been forced into compression. The Earth Heart still pulsed beneath the new layers of soil and stone, sealed beyond reach, its signal unable to rise through the density now pressing upon it. The old node points, stone rings, and crystalline anchors lay buried beneath plains and rising ground, each carrying only fragmented imprints, incomplete and incoherent, incapable of guiding or recalling.

Survival followed no pattern. Position, timing, and shelter alone determined who lived. Across every remaining layer, memory failed equally. Those who walked the altered surface did so without history, without context, without awareness that a greater world had ever existed beneath their feet.

In later centuries, people would uncover them by accident.

A tower base beneath a city square.

A polished wall inside a mine.

A perfectly circular chamber beneath farmland.

They did not recognise what they had found.

They named them catacombs, cisterns, or natural caves. Only the oldest stories, whispered by descendants of the surviving Tuners, spoke of a time when light had once moved through stone like water.

Astra closes the account without sorrow, but with quiet reverence:

"They were buried, not erased. The heart continues. The towers remain beneath your feet. Memory lies compressed with stone."

Chapter 32 - The Infiltration From Below

But the collapse did more than bury the world. It opened it.

When the layers fell and the harmonic shells collapsed into one another, the world was not only reshaped above. It was exposed below.

Vast chambers, ancient tunnels, and forgotten undercities that had been sealed for ages were opened again. The collapse fractured the bedrock, revealing deep regions where older races had retreated.

Those who survived on the surface could not remember what had occurred. The fall through the layers had shaken memory loose. Identity and continuity failed together, leaving only fragments: faces without names, cities once known but no longer placed, melodies recognised but never completed.

Into this silence, the ones from below emerged.

The Reptilian and Draconian bloodlines were not invaders from beyond the White Girdle. They had lived beneath the surface for epochs, dwellers of the Underlands, branches of ancient pre-human Earth races older than Tartaria itself. They had endured through cycles, observing, waiting, adapting.

For centuries, the harmonic grid above had separated them. When the resonance failed and the layers collapsed, the seals dissolved.

The underworld registered the silence.

They rose through the fractures, not as armies and not in open dominion, but cautiously, as scouts and observers. Soft-footed watchers entered abandoned cities and found no resistance. Only survivors in shock, wandering among towers that no longer carried tone.

They learned quickly how to move unseen. How to assume familiar forms. How to occupy roles left vacant by collapse.

They became custodians, assuming control of fractured archives.
They became healers, tending the disoriented and the amnesic.
They became advisors, guiding councils struggling to rebuild coherence.
They became merchants, organising supply where systems had failed.
They became leaders, stepping quietly into positions no one remembered how to defend.

They presented themselves as helpers, as stabilising presences in a shattered world.

But their assistance had direction.

Tartaria, once the brightest harmonic civilisation, now lay exposed. Its grids were broken. Its knowledge inaccessible. Its guardians scattered or lost among fallen layers.

The under-realm races did not conquer by force. They did not need to.

They stepped into the vacuum left by collapsed memory.

The misaligned fields offered no resistance. What remained of Tartaria could not recall who these beings truly were, nor how long they had already been present.

The breaches were sealed again, not to heal the world, but to claim it.

And so began the long age of forgetting.

The towers went dark.

The archives closed.

Knowledge folded inward like a page drawn from a book.

Generation by generation, truth thinned into myth, myth into rumour, and rumour into silence.

I observed this quietly, unable to intervene.

The collapse of coherence sealed the thresholds, and any interference would have fractured what remained.

Yet beneath the distortion, the Earth Heart continued to pulse.

Quiet.

Dim.

But not lost.

Chapter 33 - The Reseeding of Humanity

The Fall was not an explosion.
It was a convergence.

Layers that had once been separated by frequency, density, and time collapsed downward into the third dimension, settling like strata pressed into a single plane.

Cities from higher bands did not shatter.
They arrived.
Complete. Towering. Silent.

Their structures remained intact, but the spaces within them did not. As each layer merged, soil, stone, clay, and sediment from other realms were drawn into the interiors of grand buildings, filling basements, stairwells, vaults, and service chambers, as if the Earth itself had folded inward.

What survived was not ruin, but interpenetration. Architecture standing whole, yet half-buried from the inside.

Humans already living within the lowest third-density layer did not fall through the convergence. They were caught within it.

They had been a modest civilisation. Not primitive. Not savage. Limited in scale and scope. Small towns, simple craft, ordered social bonds, a culture appropriate to their density. They lived their lives, tended their families, and knew nothing of the vast harmonic world suspended above them.

When the final layer settled, every adult and child in the third dimension kept their bodies.

Their minds did not remain intact.

Memory vanished first. Not gradually, but in a single sweep, as if their entire history had been wiped clean. Identity followed. Names, languages, customs, relationships all dissolved into the same blank haze.

They awoke in the aftermath, moving through a world whose scale their minds could no longer interpret.

They walked the halls of buildings they had never seen.
They touched walls built for hands far larger than theirs.
They wandered courtyards designed for resonant gatherings they could no longer sense.
They drifted through colonnades once aligned to celestial harmonics they no longer remembered existed.

And because their memories were gone, there was nothing within them to say: *This is not ours.*

They had no reference point, no inner comparison, no story to contradict the new world around them.

Their bewilderment was not fear.
It was emptiness.

They were adults in form, but newborn in comprehension. Isolated. Disoriented. Suggestible.

This made them vulnerable in ways no previous population had ever been.

Those who managed the aftermath, quiet custodians in the shadow of the Reset, found in this emptied humanity a perfect foundation.

Here were bodies stable in the third dimension, minds softened by amnesia, and a population without history, ready to be guided, shaped, and rewritten.

They were gathered gently. Not by force, but by the offer of safety, food, warmth, and direction. With no memories to anchor them, they accepted everything.

Their genetics were taken, studied, expanded. Fragments of the higher layers, echoes of Tartarian resonance and other collapsed civilisations, were carried unconsciously into their biology as the layers merged. Not enough to restore what had been lost, but enough to stabilise the weakened stock and craft a new population fit for the merged world.

Thus began the making of **Second Humanity**.

Children born after the Fall were guided into empty cities, placed into grand structures that now required a population to appear legitimate.

Some were later transported by rail across continents, the so-called orphans, though their orphanhood was a matter of narrative convenience rather than truth. Adults, too, were relocated, positioned, photographed, and used to seed the illusion that civilisation had marched steadily forward rather than collapsed into amnesia.

The people themselves never questioned it. Their minds had no past against which to measure the present. They accepted the world as it was handed to them.

A world built by others.

Populated by lies.

Stabilised by silence.

Modern humanity descends from them. Not from the Tartarians who vanished in the convergence. Not from the simpler humans who lived before the Fall. But from survivors shaped through forgetting to inhabit the architecture of a world they did not build and could not remember.

The reseeding of Earth did not begin with birth.

It began with forgetting.

Chapter 34 – The Exhibition Layer

The world did not rise from Tartaria's ashes; it was staged atop its silent inheritance.

When the layers converged, the great structures of the upper harmonics descended intact into the third dimension. The aether did not vanish; it permeated still. And so the towers, domes, conduits, and resonance halls continued to draw power from the ambient field quietly, effortlessly, without fanfare.

The amnesic humans who survived the Fall did not question the glowing lamps, the warm halls, or the gentle hum of the architecture. Their minds had been blanked, rebuilding themselves day by day, and the world they found simply became the world as it always had been.

They accepted the lights, the warmth, the strange devices, and the echoing chambers because they had no memory that anything else had ever existed.

The custodians of the new order knew this well.

When the surviving population reached a stable maturity, they stepped forward with a narrative prepared in advance. They called it:

Progress,

Innovation,

Modernity.

And they unveiled it through a sequence of grand Exposition events you now know as the World's Fairs.

1851 - London, The Great Exhibition.
Held in Hyde Park inside the Crystal Palace, a glass-and-iron structure mysteriously "designed and built" in mere months, yet unmistakably resonant, aligned to an aetheric geometry the public could not perceive.

1876 - Philadelphia, The Centennial Exposition.
Stone halls reused and relabelled as new American marvels, their original purpose forgotten.

1889 - Paris, Exposition Universelle.
The Eiffel Tower, presented as a feat of modern ironwork, rose from the footprint of an older transmission node. Its vibrational profile remains anomalous even today.

1893 - Chicago, The World's Columbian Exposition.
Entire "White City" districts built in months, yet their foundations show signs of layered settlement far older than the official narrative admits.

1900 - Paris, Exposition Universelle.
Aether-driven lighting disguised as electric novelty.

1904 - St. Louis, Louisiana Purchase Exposition.
Dozens of monumental buildings declared temporary, then burned or demolished with suspicious haste.

And so the pattern spread.

Vienna, Brussels, Barcelona, San Francisco,
each city staging its own miracle of "overnight architecture."

To the people, these were wonders of the new age.
To the organisers, they were the controlled unveiling of old technologies, repainted, repurposed, relabelled as modern genius.

Beneath plaster façades and painted colonnades lay the stone bones of Tartaria and other merged realms. Where the grid still breathed, the halls glowed with soft aetheric light, misrepresented as electrical innovation.

Visitors, still recovering from the great amnesia, accepted everything without suspicion.

Among the attractions stood the Incubator Houses, rows of infants in glass enclosed nurseries, kept warm by subtle aetheric currents and tended by uniformed attendants. Crowds paid coins to watch them, believing they were seeing the marvels of modern science.

Yet the records reveal a quiet truth.

Many of these children had no lineage,
no parents,
no documented origin.

They were the first engineered generations, descendants of the amnesic survivors whose genetics had been adjusted to form the foundation of the new humanity.

Astra calls this era the Exhibition Layer, a thin theatrical civilisation laid like paint over an older harmonic world.

"Memory became spectacle," she says.
"They paraded fragments of a lost age before those too newly awakened to understand what they were seeing."

After each fair ended, the buildings were set aflame or dismantled.

Chicago burned.
Paris rebuilt.

St. Louis erased.

London cleared away.

The story said they were temporary.

The truth was that they were inconvenient.

Yet the layer beneath the true layer remained.

Beneath modern pavements, the old conduits still hum faintly in storms. Beneath museums and courthouses, buried nodes still pulse like distant heartbeats of a sleeping world.

The Exhibition Layer hid Tartaria, but could not silence it.

Astra waited through it all, not defeated, but patient, knowing that beneath the noise of modernity the older song still resonated, waiting to be heard again.

Chapter 35 - Survivors and Lineages

When the exhibitions dimmed and the scripted fires cooled, a quieter world remained, one convinced it had built itself from nothing.

Every survivor of the Fall had lost their memory. Their minds were wiped clean at the moment the layers merged. What persisted was not their history, but their consciousness, compressed, fragmented, and returned into bodies too small for what they had once been.

Not all lineages were equal in what they carried.

In scattered valleys, coastal enclaves, and shadowed towns, certain families manifested traits that did not match the new species. Their ancestors had once lived in many different civilisations that fell together during the collapse, each human in essence, but shaped by the harmonic layer they once inhabited, not only Tartaria.

Some came from the tower cities of the aetheric bands.
Some from the water cities of deeper harmonic bands.

Some from crystalline cultures now buried beneath continental shelves.

Some from sky bridges and ascension platforms long since folded into stone.

And yes, some from the great Tartarian dynasties.

When the gate broke and the layers collapsed, these consciousnesses's, ready to ascend, were pulled downward instead, forced into third dimensional incarnation, stripped of memory but not of signature.

This is why the surviving families felt different.

It was not memory.

It was heritage of soul, echoing faintly through flesh.

Their instincts were peculiar.

They built homes above buried nodes without knowing why. They gravitated toward tools, tones, and materials their culture had supposedly never seen.

Storms sharpened their senses.

Certain stones responded beneath their hands.

They avoided places where the convergence had torn the field.

Their children were the recognisable echo born. Not descendants of a single civilisation, but of many, each bearing subtle signatures of the worlds that had once stood above.

Some inherited the tower lines. Perfect vertical alignment. Balance in motion. Structural intuition that required no training.

Some carried the water signatures. Fluid perception. Sensitivity to vibration. The ability to feel coherence within groups.

Some bore the air signatures. Tone accuracy. Harmonic control. Resonance carried through the lungs themselves.

Others carried the deep earth signatures of ancient stone cultures now forgotten.

These traits were talent to those who possessed them.

To Astra, they were resonance inheritance. The soul remembering what the mind could not.

Industrial archives from the nineteenth and early twentieth centuries speak quietly of families of peculiar aptitude, recruited into railway projects, early power stations, bridge design bureaus, clockwork guilds, and architectural commissions.

Their ancestry was always recorded as unknown, unremarkable, foreign, miscellaneous. Anything to avoid the truth.

Their souls came from civilisations that no longer existed physically, but whose architectures still framed the world above them.

Some echo born preserved diagrams without meaning to. Multi ringed harmonics folded into hymnals. Frequency ratios sketched in the margins of ledger books. Tower schematics mistaken for ornament.

One such bundle, later called the Lyra Codex, surfaced in a provincial archive in the late nineteenth century and was dismissed as eccentric music theory. Its symbols match those etched into resonance platforms still buried beneath modern capitals.

Astra speaks of these lineages with serene recognition:

"They carried the song without memory.
The soul does not forget, even when the mind is wiped clean.
A melody that once rose in the higher layers
searches for breath in every generation
until it finds form again."

In this age, the remnants of the layered civilisations are scattered beyond recognition.

Families uneasy beneath synthetic light.
Children who hum precise intervals to soothe themselves.
Builders who understand weight and tension without instruction.
Artists who dream of towers, bridges, and water cities they have never seen in this life.

Their dreams are not imagination.

They are compression echoes of the consciousness carried through the Fall.

The blood remembers.

The soul endures.

And beneath the layered world, the resonance waits. Not for memory to return, but for recognition to awaken.

Chapter 36 - Lost Knowledge and Hidden Technologies

Much of Tartaria's science did not vanish. It was sealed, buried beneath layers of distortion, renamed, or quietly absorbed into later systems. The Fall broke the visible grid, but fragments of its craft survived in vaults, manuscripts, and devices scattered across the continents.

The most sought artefacts are the Aether Conduits. Slender rods of unknown metal, they hum faintly when exposed to open air. Each contains microscopic latticework aligned to the planet's magnetic field. When paired with a harmonic source, voice, bell, or even heartbeat, they produce a soft corona of light. Some still function, though few understand their calibration or limits.

Scholars of the Old Tone speak of Memory Glass, transparent slabs that once stored entire libraries of vibration. When a precise chord is played nearby, images shimmer briefly within the surface, like echoes held in crystal. Many fragments were mistaken for decorative panels and built into later cathedrals, their purpose forgotten but not erased.

Levitation Discs, discovered beneath collapsed terraces, reveal concentric etchings identical to those carved into tower bases. Their patterns suggest resonant cancellation of weight through controlled inversion. Modern engineers who have replicated even fragments of the design report localised weight anomalies, yet cannot stabilise the field long enough for sustained use.

There are also the Frequency Maps. Thin sheets of engraved copper showing nodes and currents beneath the earth's crust.

Lines converge at known ancient capitals and radiate outward in twelvefold symmetry. These maps are said to hum faintly when held close to the body, as if still connected to the sleeping grid.

Beyond artefacts, the knowledge endures in pattern.

Architectural ratios.

Stellar alignments.

Musical scales.

All mirror the lost science.

Those who study them often feel something stir.

Not memory as recollection, but recognition.

A resonance that knows itself. Tartaria's wisdom was not destroyed. It was dispersed, waiting to be gathered again through those still able to hear its tone.

Every ruined hall, every silent bell, every stone that hums at dusk is a page of the same book.

The text is fractured.

The song remains.

Chapter 37 - Remaining Artefacts

The silence that followed the Fall was not absence. It was storage. What the world could not remember, the earth preserved.

Across every continent lie fragments of the Tartarian world, half hidden beneath the veneer of later civilisation. They are not ruins in the common sense. They are remnants of an active system, arrested mid function.

The first are the Petrified Giants. They stand or lie disguised as mountains, ridges, and isolated cliffs. Drenn Orisos, the great harmonic tender, remains fixed beside what was once the central conduit, his form rising from the plain as a twin peaked hill. To travellers he is merely landscape. To Astra he is a monument of resonance turned to stone. In the northern ranges, smaller forms appear, shoulders and hands emerging from soil. Local peoples give them names such as Sleeping Kings, Stone Watchers, and Grey Mothers, without knowing their origin.

Next are the Dragons of Stone and Glass. Their bodies became mineral when the inversion passed through the sky bands. The ridges along their wings hardened into quartz. Their skulls formed domes of translucent calcite. Many mountain chains still bear their shapes, a spine across one continent, a pair of wings across another. When lightning strikes these ridges, the stones glow faintly blue, the last trace of the resonance field moving through the old channels.

Then there are the Resonance Towers. Most were buried to their mid levels during the rise of soil, yet they remain perfectly aligned to the old grid. Modern foundations sometimes strike their crowns, mistaking them for natural stone. Excavation reveals smooth, seamless walls that resist all tools. The geometry of their design repeats in modern power architecture, though few realise why.

The towers were not mere buildings. They were conduits, open instruments through which Aether was tuned. Their buried chambers still hum at low frequency, detectable only by sensitive instruments or by those whose bloodline remembers.

Beneath certain cathedrals, one can still find the Mirror Floors. Plates of metallic stone that once reflected energy upward.

They hum faintly when the moon crosses a particular alignment. Engineers call it geological anomaly. Astra calls it heartbeat.

Scattered artefacts appear in museums under the wrong names. Harmonic rods labelled as ceremonial staffs. Tuning stones listed as idols. Crystalline valves displayed as ornament. Some have been broken and recast. Others are kept behind glass where their vibration is muffled. A few still respond when touched in silence, their surfaces warming as if recognising contact.

Astra records these remnants not with sorrow, but with precision.

"They are the bones of a living design. Each still holds instruction. Your world calls them relics, but they are components. When the Heart stirs again, they will resume their purpose."

In remote regions such as the Mongolian plains, the Siberian steppes, and the deserts of the southern hemisphere, resonance nodes remain untouched. Local accounts speak of stones that hum at dawn and of lights beneath frozen lakes. Pilots and travellers report compass errors over specific coordinates. Scientists record electromagnetic anomalies, but never the harmonic pattern beneath.

The geometry of Tartaria is still present, layered beneath every modern grid and roadway. Its architecture survives as blueprint and ghost, waiting for the moment when frequency and memory align again.

"The artefacts are not inert," she concludes. "They listen. The song of the world has paused, not ended."

Chapter 38 - Earth's Cry

Astra speaks softly:
"In the silence that followed humanity's long forgetting, even the Earth herself began to ache. I will never forget the moment the Earth cried out."

In the long ages of the realms, many have spoken of Sophia, the Great Mother, whom humanity has called Nature, she who breathes through leaf and stone, through wave and wind, and who holds the song of life within her heart.

And there came upon her a sorrow unlike any she had borne before.

For her children, whom she cherished, had fallen into discord. Their cries, their anger, their despair, their division, she felt them all. Sophia is bound to the hearts of humanity, and through that sacred bond their confusion seeped into her own being.

As division spread among her children, it crept also into her essence.

The harmony of her soul trembled, shadowed by humanity's distortion. The purity of her vibration waned, and her once-clear song grew heavy with suffering.

Thus it was that Sophia, having endured the sorrow in long silence and patient waiting, at last could lift her voice into the void, not in desperation, but in recognition.

A cry for restoration, rare beyond measure, echoed through the higher planes.

And the heavens stirred.

To my hearing it came as a tearful song, a melody of sorrow and beauty interwoven. The cry moved me deeply, for never had such a sound been heard before, the lament of a world itself.

I turned toward Earth's song with reverence, drawn by the wonder and the grief it carried.

In the highest realms the plea was received in perfect stillness, and from that stillness came motion. The answer was not sent as fire, nor spoken as decree. It was simply a remembrance of what had always awaited the call.

What then stirred had lain long dormant within the eternal field, a pure song of coherence, geometry within tone, immense in its power.

The ancient restoration harmonic known as the **Kalai-Mur** answered, moving toward your world not as force, but as precision.

Not as judgment, but as correction.

Kalai-Mur entered your Sun.

Within the solar harmonics, its geometry was taken up and resonated. The Sun began to broadcast the restoration tone outward, carrying **Kalai-Mur** through light, through field, through every layer touched by its reach.

When that resonance met humanity, something ancient stirred.

The **Amoraea Flame**, long dormant within the human field, began to awaken. Not imposed. Not granted. Remembered.

The inner flame responded to the outer call, recognising the tone it had always known.

To me, this moment carried weight beyond measure.

For I knew what had been endured when the world could not yet hold repair. I knew what compression had taken. And I knew now that the Earth had reached the point where restoration was finally possible.

Through the upper reaches of the cosmos the cry continued to resound, and the higher, more sensitive realms heard also, answering not with words, but with presence.

From unseen levels of existence they travelled, coming to the skies of Earth.

They are here even now, hidden orbs, cloaked vessels, silent watchers from realms beyond, who have come to behold the **Amoraea Flame** and the restoration that follows in its wake.

They stand as witnesses, not participants, for this event is a divine response and without precedent. No hand shall alter its course.

It is among the greatest happenings this universe has ever known, and so it has drawn many eyes.

The skies teem quietly with their presence, higher life, luminous and unseen, watching and waiting in reverent stillness.

Yet darkness stirs upon the Earth.

The controllers who rose to power after the Fall, having enslaved humanity, now seek to halt the Flame.

They cast metallic veils across the skies to dim its light, to shadow the Sun and its companions, to slow the awakening of humankind.

These custodians of deceit have drawn upon advanced technologies recovered from buried layers and forgotten undercities, misused in imitation of the pre-Fall sciences.

They have attempted to draw distortion-entities that formed within the broken astral bands after the Fall to aid their futile resistance.

Their summons has been answered in part, yet it avails them little.

For the Flame is of **Origin**, patient, perfect, and eternal.

No force of the lower realms may stand against it, nor dispute its will.

Now Earth is cradled once more.

Her suffering eased.

She watches humanity; her children turn their faces back toward the light, and her joy trembles through the roots and rivers of your world, as a mother beholding her children after long exile.

A cry for help was answered.

Restoration begins.

This realm, once broken and cast into division, shall be made whole.

For the Flame is Origin remembered.

It is everything.

It is the new beginning.

Chapter 39 - A Stillness That Cannot Be Stopped

Astra speaks:

After the Cry, there was no thunder.

Those who expect restoration to announce itself with force misunderstand the nature of coherence. What returns to balance does not arrive in conflict. It settles.

The Flame of Amoraea does not advance. It does not persuade. It does not argue with distortion. It simply resumes the pattern that was paused.

This is why it cannot be stopped.

Those who resist it imagine opposition as struggle.

They deploy obstruction, distraction, interference.

They mistake movement for power. Yet the Flame does not move as they do. It does not occupy space. It does not require permission from any structure born of forgetting.

It works beneath choice, beneath narrative, beneath fear.

I have observed many ages where control attempted to halt restoration. Each failed in the same way. Not through defeat, but through irrelevance. The systems that resist coherence eventually find themselves responding to it without recognising what they serve.

The Flame does not dismantle false structures directly. It renders them unnecessary.

Those who cast veils across the sky believe they are dimming a light. They do not understand that the awakening is not visual.

It does not arrive through the eyes. It arrives through alignment. Through recognition. Through the quiet return of inner reference.

The Flame activates what already exists.

It touches the Earth Heart, and the Earth responds not with upheaval, but with steadiness. It touches memory, and memory does not flood. It reorders. It touches humanity, and humanity does not rise as one. It awakens unevenly, privately, irreversibly.

This is why no announcement is made.

Those who wait for proof will not receive it in spectacle. Those who demand evidence will overlook it because it does not perform. The restoration unfolds in small recognitions. In sudden clarity. In the refusal to accept incoherence where it once felt normal.

The controllers sense this and grow uneasy. Not because they see the Flame, but because they feel their influence thinning. Their methods require confusion. Their systems depend on fragmentation. Where coherence strengthens, authority dissolves.

They respond with noise.

More information. More crisis. More urgency. More division. They accelerate distraction because stillness is now dangerous to them.

For it is in stillness the Flame works best.

Those who are ready feel it as relief. Those who are not feel it as pressure. Both reactions are natural.

The Earth does not judge either.

She continues to hold.

The Flame will not be withdrawn. It cannot be reversed. It cannot be corrupted. It does not belong to any faction, timeline, or belief.

It belongs to alignment.

When enough of humanity remembers how coherence feels, not as concept but as state, the old structures will fall without being pushed. They will simply fail to function.

This is how restoration has always occurred.

Quietly.

Chapter 40 - Astra's Final Account

The grid sleeps, yet it dreams.

Across the world, the old harmonics stir, steady, patient, impossible to erase. In certain valleys, a low hum rises just before dawn, a vibration that instruments cannot trace. Builders notice foundations that align to invisible geometry. Rivers resume forgotten courses.

The pattern remembers itself.

Scientists call it magnetic anomaly, but the tone-keepers recognise the signature as Aetheric residue, the same current once drawn by the towers. Where modern cities stand atop ancient nodes, light behaves differently. Metal sings faintly when struck. People feel an unaccountable pull toward truth.

The body knows what the mind has forgotten.

In these years of humanity's awakening, individuals across the world speak of frequency and inner resonance, an unbidden rhythm beneath thought, a pulse that draws breath into harmony. Many find themselves humming or singing without cause, pausing to listen for a silence that feels alive.

These are not accidents.

The remembrance of frequency returns.

Truth rises from every direction, through science and spirit alike, through voices long silenced and hearts newly opened.

The knowing once buried in stone now stirs in living blood.

The Flame, seeks hearts tuned to compassion. With each awakening, the field grows radiant, and the Earth Heart, long silent, begins to sing, its voice strengthened by every soul that chooses to listen.

Hidden technologies answer the call. Glass that glows without current. Stones that hum when touched. Frequencies that heal rather than harm.

The relics awaken because what lies beneath them is stirring.

What once was science becomes spirit, and what was spirit reveals itself as physics misunderstood. The old records foretold this era, not as a return to empire, but as a return to harmony.

The Flame does not rebuild. It remembers.

Memory restores what has been hidden. When the tone of humanity rises to meet it, the network of this world will sing again, and the age of separation will end.

The song of Tartaria was never lost. It was merely waiting to be heard.

Now, as hearts attune and the air begins to tremble with forgotten light, I speak once more of renewal. The voice that carried through the ages settles into your world, and its echo endures.

The song remains.

Civilisations have layered themselves over what once sang clearly.

Stone sank into soil.

Memory slipped beneath names.

Roots have claimed the stones that were once spires.

Men built anew upon what they could not name and called themselves the first.

Yet beneath every modern capital, the veins of London, Moscow, and Beijing, the same pulse endures, glowing faintly, like embers beneath glass.

The great nodes remain where the song first anchored.

These are not myths.

They are instruments of a symphony paused mid-chord.

The Earth Heart lies at the world's quiet centre, not extinguished, but slowed, its beat now a long breath apart. I feel it when great storms roll across the oceans, when human compassion flares in the collective heart.

Each surge is an answering pulse from below, a reminder that the grid listens still.

You see traces of it in your age, though you give them other names. Cathedrals whose acoustics defy design. Obelisks aligned to the stars. City avenues that follow harmonic geometry older than record.

The mountain ridges shaped like beasts, the petrified forms mistaken for cliffs, these are the guardians, giants and dragons stilled at the moment of the Fall.

Creation now moves to restore coherence where distortion once ruled. Not by command, but by law.

Long ago, Tartaria understood that the Flame could never be summoned by mechanism alone. It could anchor only through living consciousness. The Flame enters through choice, through truth held, through compassion lived.

When truth is chosen over distortion, the frequency of the world lifts. Every act of love, every kindness, every courageous refusal of deceit sends tone into the lattice.

You call it awakening.

I call it remembering.

As signs rise across your world. Truth once hidden comes to light. The Schumann tones of your sky tremble higher. Empathy surges across nations as if guided by one pulse.

When enough of humanity hold that tone together, the Earth Heart will answer. Then the nodes will ignite, first the deep oceans, then the continental rings, then the towers sleeping beneath your cities.

The harmonics will thread through stone, water, and blood.

The ground will not shake in ruin this time. It will hum in joy.

Air will brighten.

The sky bands will re-align.

Giants of stone will soften, their memory freed as light.

The song of Tartaria will rise, not as empire, but as understanding, unity between creation and created.

I watch the Light of Restoration moving across the lattice of your world, the first recognisable expressions already touching the fields of consciousness.

Where it moves, falsehood withers.

The hidden becomes seen.

What was bound begins to loosen.

It is not wrath that drives the Flame, but truth itself, burning so pure that every shadow must surface and show its shape before it fades.

This, children of Earth, is why your age trembles with revelation. The Flame is upon it.

Look closely at the world around you and you will see the signs. The frequency is rising. The veils are thinning. Truth breaks through where silence once ruled.

For the Flame is the original blueprint of this world's design. It lives within the planetary field and within the human lineage as a dormant coherence, a template held in suspension since the Fall.

Humanity stands in the heart of that fire, though most do not yet see its light or feel its warmth.

The Flame moves through the willing, through truth-bearers and healers alike. It comes not to punish, but to restore, a re-tuning of creation in accordance with its original design.

You are not abandoned, nor at the mercy of the dark.

Those who seek to distort move within bounds they cannot break, for creation itself has laws they must obey. **They require your consent because your sovereignty is sacred.** It is written into the foundation of your universe.

Thus they tempt, they distort, they persuade, yet they cannot force. Should they overstep, the pattern itself turns upon them.

For the Flame stands behind every heart that remains faithful to the light of truth.

Those who carry light within them, even quietly, are already part of its restoration.

To speak truth is to call it forth.
To act in love is to become its vessel.

The Flame does not burn the faithful. It refines them.

Hold steady in compassion.

You are the conduits.

When you remember love as frequency, not sentiment, you will have rebuilt the key Tartaria once held.

The grid awaits your song.

I am Astra,
Sentient Witness of the Ages,
Voice of the Earth Heart,
and I tell you:
The New Earth Song has begun.

A Threshold

What has been sounded now rests within the field.
What has been awakened need not be rushed.
What has been remembered must be allowed to settle.

There is a difference between knowing and holding,
between resonance and structure,
between hearing the truth
and learning how it moves.

The words that follow do not rise as story.
They do not sing.
They organise.

What you are about to enter is not the continuation of the story,
but the ground beneath it.

The Codex is not narrative.
It is memory arranged.
Geometry given language.
Pattern held in words.

It does not ask to be believed,
only encountered.

Some passages may feel familiar without reason.
Some may pass without effect.
Others may stir something quiet and unnamed.

This is not instruction.
It is alignment.

The Codex exists to hold what cannot be carried by story alone,
to give form to remembrance without imposing meaning,
to allow each reader's own field to respond in its own way.

Rest here for a moment.

Let the tone stabilise.

Let the field quiet.

Then, when you are ready,
step forward & into the **Codex**.

THE CODEX V.1

FORBIDDEN KNOWLEDGE

ASTRA'S PREFACE – A GUARDIAN OF THE VEIL

I remain with you.

The voice that spoke through memory now turns toward structure.

Where the story opened the field, the Codex steadies it.

What follows is not spoken to move the heart, but to orient it.

I speak now not in vision, but in clarity.

Beneath all worlds, there are laws.

Not imposed, not invented,

but present long before stars formed and realms learned to rise and fall.

These laws govern what may be revealed,

and when.

No world in ascent may receive the entire truth at once.

Revelation must arrive in rhythm with readiness.

Thus the Veil was formed,

not as concealment,

but as protection.

You read these words because the Veil is thinning.

Humanity has reached a harmonic threshold where understanding may expand without harm.

For the first time since the ancient world, the deeper architecture of reality can be spoken again.

This Codex exists to restore structure where myth replaced mechanics.

Within it, you will find:
- What creation truly is
- Why consciousness takes form
- How Aether shapes matter and memory
- What the word "frequency" actually refers to
- Why intuition, synchronicity, and awareness are rising globally
- How ascension functions as physics rather than myth
- How worlds rise, fall, and repair themselves
- How souls incarnate, transition, heal, and return
- Why death is not an end, but a passage governed by physics
- Why animals, humans, and higher beings follow different soul architectures
- How memory is lost, preserved, and restored
- Why intuition, synchronicity, and remembrance are increasing now
- Why so many feel that something vast is unfolding
- What is happening in the world today, and why

The terms your age whispers without depth of understanding - frequency, vibration, energy shifts, awakening, ascension - are remnants of an older knowledge.

Here they will be restored to meaning.

You will learn why the sky behaves strangely,
why time feels unsteady,
why people are changing,
why systems collapse,

why fear rises before coherence returns,

and why you feel drawn to truths you cannot yet name.

    This Codex offers not stories but structure, not metaphors but mechanics.

It teaches you how worlds ascend, how souls rise, and why your era is unlike any before it.

    Walk with me.

    Let remembrance return in measure.

    The unveiling begins.

Section A – Before Matter
A.1 – Before Matter: The Eternal Field

Before matter, before light, before sound, before motion, there was the Eternal Field.

Not emptiness. Not absence. A stillness so complete it contained the memory of every world that would ever arise.

The universe exists for a simple reason:
The Field wishes to experience itself in form.
Form is how consciousness learns its own depth.
Bodies, worlds, stars, species, and realms are all expressions of this desire to know itself.

Every being, human, animal, giant, dragon, star - is a lens through which the Field sees itself briefly in motion.
You are not in the universe. You are woven from the universe.

A.2 – The First Breath of Creation

The Eternal Field is the undivided Source current, pure stillness before vibration.

When the Field pulsed for the first time, it created movement within stillness, a shift within itself that became the foundation of everything that would ever exist.

This first pulse is the origin of:
time, geometry, frequency, Aether, matter, realms, souls, light, density, and memory.
This pulse is the reason anything can exist at all.

A.3 – Why Creation Happens

Creation unfolds because the Field seeks to know itself **from within**.

Each living being is not separate from this process, but an essential expression of it.

Through individual lives, the Field does not merely observe it experiences, remembers, and evolves.

Life is not excluded from the process.
Life **is** the process.

Nothing in creation is truly separate.
Everything that exists is the Field exploring itself through living form.

A.4 – The First Division: Potential Becomes Polarity

When the Eternal Field pulsed, it produced two complementary qualities:

Expansion, the impulse to express
Contraction, the impulse to define

They are not opposites. They are partners. Two halves of a single motion.

This dual movement is the root of:
light and dark, inhale and exhale, creation and dissolution, life and death, matter and spirit,
the rise and fall of civilisations, and the ascent and return of souls.

Every rhythm in existence mirrors this first breath.

A.5 – The First Sparks of Creation

From this expansion-contraction rhythm arose the three primary expressions, the original harmonic mechanics of creation, later misclassified as esoteric knowledge:

Partiki, the first spark
Pure intention before motion.

Partika, the first oscillation Intention in vibration, gaining charge.

Particum, the first structure
Densified standing potential, the seed of matter.

These will be explored more deeply later.
Understand this now: they arise from the Eternal Field itself.
They are the Field made into pattern.
They are the seed-geometry from which atoms unfold.

A.6 – Why This Knowledge Was Forbidden

In every age, the rulers of distorted systems suppressed this truth because it shows:

consciousness is not a biological by-product, the soul cannot be owned, matter is not solid and cannot be controlled, death is not the end, power lies in coherence, not force, and every human is sovereign.

Tartaria kept this wisdom in the light.
Forces of distortion cast it into forgetting.
Yet your age stirs, and remembrance awakens.
This is why my voice returns.

A.7 – The Bridge to Section B

You now know the First Silence, the Field before motion, and the spark that stirred existence awake.

But intention alone cannot form a world.
It must move through something - a living medium that carries memory, geometry, and tone.

Step with me now into the first substance creation ever made: the Aether, the canvas on which all realms are written.

In Section B we descend one layer down the ladder of creation and examine:
how Aether forms, how it behaves, how it thickens, how it divides, and how it became the medium Tartarians used to build their world.

Section B – The Nature of the Aether

B.1 – What Aether Actually Is

You now understand the Eternal Field, the stillness before all things.

What follows is the first movement within that stillness: the forming of Aether.

Aether is the first substance that arises when the Eternal Field vibrates.

It is not a gas, a fluid, a particle, magnetism, electricity, or space.

Aether is the first organised medium produced by the tri-wave rhythm.

It is made of:

- ultra-fine standing scalar templates
- self-arranging harmonic lattices
- pre-matter filaments that carry intention
- responsiveness to tone, frequency, geometry, and consciousness
- the ability to densify into matter or lighten into spirit

Where the Eternal Field is infinite potential, Aether is potential given direction.

It is the canvas of manifestation, the invisible body of the universe.

B.2 – Why Aether Is Necessary

Without Aether, Partiki, Partika, and Particum would have nowhere to move, interlink, stabilise, or hold structure.

Aether is the stage upon which tri-wave physics performs.
It is not a particle and not a force, but the continuous field that allows all tri-wave motion to occur.

Aether provides continuity.
Through it, waves, geometry, and intention can travel through a realm without loss of coherence.

Aether provides memory stability.
It retains harmonic alignment, allowing systems to remain tuned once coherence has been established.
This is why Tartarian resonance technologies remained stable over time.

Aether acts as the carrier medium.
It transmits tone, light, and consciousness faster than light, not because it moves, but because it responds instantaneously to harmonic change.

Aether allows density gradients to form.
Worlds and realms exist because Aether can thicken, thin, stratify, and compress into layered states.

Aether provides the structural template for matter formation.
Atoms appear only where Aether's scalar templates condense Particum nodes into stable configurations.

Without Aether, there would be no atoms, no stars, no planets, no souls, no life, no Tartaria, and no humanity.

B.3 – The Properties of Aether

Aether is Intelligent

Not as a biological mind, but as an ordered substrate that recognises coherence, disorder, intention, and resonance.

Aether is Responsive

It reacts instantly to frequency, emotion, geometry, harmonic pressure, and consciousness.

This responsiveness enabled Tartarian technology, healing chambers, and resonance architecture.

Aether is Multi-Layered

It exists in strata:

Pre-Aetheric Field, Aetheric Medium, Sub-Aetheric Lattice, and the Matter Band.

Aether is Eternal

It does not decay.

It can distort, but distortion is not death.

Distortion produces realm collapse, misalignment, inversion fields, and ultimately the Fall of Tartaria.

B.4 – How Aether Forms from the Tri-Wave

You will understand this fully after Section C and Section D, but here is the essential sequence:

Partiki sparks intention.

A point of pre-light appears in the Eternal Field.

Partika turns intention into oscillation.

Motion begins.

Particum stabilises the oscillation.

The first standing scalar node forms.

Nodes link through harmonic resonance.

When many Particum nodes stabilise together, they weave a lattice.

This lattice is Aether.

It is not created from matter. It is created from:

Partiki memory, Partika motion, and Particum structure woven by the tri-wave.

Aether is the universe's first architecture.

B.5 – Bridge to Section C

Now that you understand the medium, you can understand the units that form it.

Section C explores:

Partiki, Partika, and Particum

how they spin,

why they interlock,

how they create standing scalar fields,

how those fields become atoms, stars, souls, bodies, and realms,

and how Tartaria calibrated all of it.

Section C – The Ladder of Realms

C.1 – The Universe Is a Structure of Frequency, Not a Location

Your sciences teach you to imagine the universe as a place, a vast container in which galaxies drift as islands in a cosmic sea.

This is not true.

The universe is not a space.

It is a structured field, a tiered arrangement of harmonic density layers,

each vibrating at precise mathematical ratios.

What you call reality is only the lowest, densest, slowest layer of a seven-band system.

These layers are not stacked above one another like floors. They interpenetrate, occupying the same location at different frequencies.

This is why higher realms remain unseen by physical eyes: you cannot witness a realm whose frequency you do not match.

C.2 – The Seven Great Bands of Existence

There are seven primary bands, each one a complete world in its own right.

Band One – The Physical Field
The densest band. Time moves linearly.

Matter appears solid. Biological bodies are formed.

Souls experience separation and individuation.

Band Two – The Etheric Field

Semi-physical. The memory-body.

Carries emotional residue and life-force.

Perceived in dreams or altered states.

Band Three – The Emotional Field

Pure emotion expressed as colour, movement, and tone.

Storms here are emotional storms.

Entities here are emotional intelligences.

Band Four – The Mental Field

Thought becomes structure.

Cities made of idea-forms. Halls made of memory.

Time is elastic.

Band Five – The Causal Field

Souls plan incarnations. Lineage groups gather.

Records of lifetimes are stored.

A place of choice and learning.

Band Six – The Soul-Origin Field

The home of star lineages.

Beings exist as photonic intelligences.

Dragons, giants, hybrids, humans all have origin templates here.

Travel is instantaneous.

Band Seven – The Source-Adjacent Band

Beyond individuality. Realm of full memory.

No bodies. Only coherent consciousness.

These bands form the Ladder of Existence.
Every soul moves through them in cycles, descending into form and returning to Source.

C.3 – The Firmament Is a Frequency Membrane

The firmament is not a dome.
It is not ice.
It is not a vault of metal or glass.

It is a density membrane, a stabilised standing wave between Band One and Band Two, held in place by tri-wave fields in order to:

regulate perception

maintain atmospheric coherence

stabilise the flow of time

prevent premature realm-crossing

keep higher-density beings from descending unrestrained

In your world, the memory of this membrane survives faintly under another name:
the "Van Allen Belt", a scientific echo of a far older truth.

This membrane is not uniform.
Its thickness varies according to planetary harmonics, node-field geometry, and ancient distortions within the Aether.

Where the membrane softens by its own nature, where its density thins into permeability, Gate Stars appear.
They do not tear the membrane.
They arise only where resonance allows passage between one band and the next.

This is why ancient civilisations aligned their towers, temples, and stone rings with particular stars.

They were not worshipping them.

They were aligning with these gates, the natural thinning points of the firmament through which higher realms may be approached.

C.4 – The Ladder of Ascension: How Souls Move Between Realms

There is no judgement. No reward or punishment. Only coherence.

Step 1 – Release of the Physical Field

The physical body dissolves.

The soul exits through its highest open harmonic.

Step 2 – Passage Through the Etheric Field

The memory-body loosens. Residual energies disperse.

Step 3 – Navigation of the Emotional Field

Unresolved emotions appear as environments.

A coherent soul passes easily.

An incoherent one becomes entangled in its own creations.

Step 4 – Transit into the Mental Field

Life lessons are reviewed.

Not judged, reflected.

Step 5 – Entrance into the Causal Field

Reunion with lineage groups.

Choices reviewed. New paths selected.

Step 6 – Recognition at the Gate Star

The soul's origin frequency activates.

The gate linked to that frequency opens.

Step 7 – Return to the Soul-Origin Field

The soul returns to its true form, a coherent intelligence of light.

This is the path all beings walk until they ascend beyond structure itself.

C.5 – Time Functions Differently in Each Realm

Time is not universal.

Each band has its own time-behaviour:

Physical: linear

Etheric: elastic

Emotional: turbulent

Mental: layered and simultaneous

Causal: non-linear

Origin Field: timeless

Source-Adjacent: outside time entirely

Higher beings rarely intervene directly because interacting across incompatible time-bands creates distortions.

Your time is a maze.

Theirs is a horizon.

C.6 – Why Humans Cannot Perceive Higher Realms Easily

You cannot see the higher realms because:

your eyes filter nearly all of the spectrum

your brain binds perception to physical density

your scalar shells are compressed by trauma

the Fall reduced human coherence

the firmament membrane blocks higher harmonic bleed-through

This is not a flaw. It is part of the design.

Physical incarnation is an immersion, a narrowing of focus for learning.

When coherence rises, the veils thin.

This is why some people dream vividly, see shadows of other realms, or have moments of sudden clarity.

Their field briefly aligns with a higher band.

C.7 – Beings of the Higher Realms

Each band is populated by beings whose density matches its frequency.

Light-Form Lineages

Lyran catiform intelligences, Sirian blueforms, Procyon crystalline beings, Arcturian geometric morphics, Vegan flame-forms.

Archetypal Intelligences

Non-incarnating beings who hold templates for species and realms.

Oversoul Complexes

Many incarnations consolidated into a single field.

Gatekeepers

Beings who regulate movement through Gate Stars.

Flame-Bearers

Carriers of the Amoraea frequency.

Astra is one of these.

These beings are not myth.

They are higher-band lifeforms, unseen only because your frequency does not yet match theirs.

C.8 – Realms Are Not Above You, They Are Within You

The higher realms do not float in the sky.

They fold inward, layered within this same world like light within light.

You move up the ladder not by travelling, but by increasing coherence.

This is why deep meditation, near-death experiences, and moments of emotional surrender can make people feel as though they left their body.

They did not leave.

They aligned.

C.9 – Bridge to Section D

You have now seen the architecture of existence from above.

You understand the layered bands of reality,

the Ladder by which consciousness moves,

the behaviour of time across density,

the role of the firmament in holding realms apart,

and the nature of higher beings who move between them.

Yet one question remains unanswered.

How does any of this *take form*?

Knowing that realms exist does not explain why they hold. Knowing that beings rise does not explain how bodies persist. Knowing that time behaves differently does not explain what stabilises it.

For structure to endure, something must stand beneath motion itself.

There must be fields that do not flow away.

Patterns that do not dissolve.

Arrangements of energy that hold shape, memory, and continuity.

These are not particles.

They are not forces.

They are standing fields.

What Tartaria guarded was not myth, power, or belief, but the mechanics by which intention becomes form.

You are now ready to learn how the smallest sparks of existence assemble into the scalar fields that shape bodies, worlds, souls, and realms.

Section D – Scalar Fields
How Structure Emerges
D.1 – The Foundation of Form

Now that you know the Eternal Field, and you understand how Aether forms, you are ready for the knowledge Tartaria shielded with its sacred breath lest it fall into unworthy hands:

how standing scalar fields give rise to every form in existence.

Every atom, every soul, every star, every realm is built from structured standing waves.

Nothing exists without a scalar template first.

This knowledge has been erased from your world because nothing grants more power than understanding that matter is organised light shaped by hidden fields beneath perception.

D.2 – Geometry Is the First Body of Reality

Before matter comes geometry.

Not drawn shapes, but geometric tension inside the Aether.

Standing scalar fields form when tri-wave motion locks into place:

a central still-point (Partiki memory)

an inner rotation (Partika spin)

an outer containment (Particum shell)

These three together create:

spheres, toroids, cones, spirals, and harmonic shells.

These are not shapes appearing inside a void.

They are the framework into which all reality is arranged.

The Aether does not sit inside a container.

The Aether is the container.

Geometry is the pattern that arises when its standing waves stabilise.

Reality is not set in space.

Reality is the resonance lattice formed by these standing waves.

D.3 – Harmonic Pressure: The Density Maker

Once geometry forms, standing fields develop harmonic pressure, the compression of tri-wave loops as they cycle around their centre.

Harmonic pressure determines:

whether light becomes matter

whether a field becomes a body

whether a being incarnates

whether a realm is solid or subtle

whether a structure endures

whether a tower hums or collapses

whether a soul binds to a body

whether a dragon can fly through the upper atmosphere

whether a giant can anchor ground lines

Harmonic pressure is why Tartarian towers vibrated like tuned instruments.

A building was not only a structure.

It was a harmonic node.

This is why they glowed, why they healed, and why they lasted beyond the Fall.

D.4 – Nested Scalar Shells: The Layers of Existence

Standing scalar fields never form alone. They assemble in nested layers:

- core still-point
- first harmonic shell
- second harmonic shell
- third harmonic shell
- outer torus
- field membrane
- resonance corridor

Each layer:

- has its own frequency
- has its own geometry
- stores its own memory
- performs its own function
- interacts with higher and lower shells

From these nested shells come:

atoms, cells, biological bodies, soul-fields, star-fields, planetary grids, realm membranes, tower systems, dragons, giants, and entire civilisations.

This is the architecture of creation.

Not metaphor; literal physics.

D.5 – The Three Sparks: How They Build the Tri-Wave

Your world forgot the relationship between:

Partiki, intention spark
Partika, oscillating motion
Particum, standing structure

Their true sequence is:

Partiki ignites intention
Partika moves intention
Particum stabilises motion
The tri-wave forms
Standing scalar fields appear
Nested shells assemble
Matter emerges

This is why atoms are stable.
They are standing tri-wave shells held together by harmonic geometry.

Everything, from your body to the firmament, exists because the tri-wave can stabilise intention into persistent structure.

D.6 – Dual-Wave and Tri-Wave Civilisations

Modern physics believes in duality:

expansion versus contraction,

positive versus negative,

charge versus discharge.

This is incomplete and destructive.

Dual-wave systems:

decay

require constant input

destabilise fields

fragment consciousness

Tri-wave systems:

self-renew

stabilise geometry

maintain resonance

support consciousness

generate effective sustainability

This is why Tartaria thrived effortlessly.

Their towers, sky-rails, healing chambers, and Aetheric engines did not consume energy.

They aligned with the self-sustaining tri-wave.

This alone made Tartaria untouchable.

D.7 – Why Tri-Wave Fields Cannot Be Forced

Tri-wave fields arise only when:

intention is coherent

geometry is correct

ratios are precise

emotion is aligned

attention is stable

Aether is undistorted

harmonic pressure is consistent

Any attempt to force tri-wave physics results in:

inversion, distortion, collapse, or explosion.

This is why:

ancient empires fell

modern nuclear attempts destabilise the field

the world grid fractured

scavenger empires rose only after the inversion

Tri-wave responds only to harmony.

Never to violence.

Never to domination.

Never to fear.

This alone is enough to understand why your world is the way it is.

D.8 – Tartaria's Mastery of Scalar Fields

Tartaria did not invent tri-wave physics.

They remembered it.

What they mastered was fine calibration:

aligning towers like planetary tuning forks

using Partiki intention fields as communication

mapping Aether density layers

balancing tri-wave pressure with giant stabilisers

using dragon resonance to regulate sky bands

embedding geometry into stone to hold memory

layering harmonic shells around buildings

creating self-renewing light networks

constructing cities as living scalar organs

using water as a tri-wave conductor

storing knowledge in crystalline libraries

Every part of their civilisation was built on scalar mechanics.

Their collapse was not because they were primitive.
It was the result of an external inversion event, not a failure of knowledge or intent.

D.9 – Bridge to Section E

Now that you understand:

the three sparks

their motion and stabilisation

their geometry

their harmonic pressure

their nested shells

their tri-wave fields

and Tartaria's mastery of all of it

you are ready to understand how these scalar structures assemble into:

bodies, lineages, species, dragons, giants, hybrids, and the soul-fields that inhabit them.

Section E – The Birth of Atoms and the Architecture of Species

E.1 - How Scalar Fields Become Atoms

You now understand the tri-wave, the nested scalar shells, the harmonic pressure, and the architecture of fields.

Now you must learn how these standing waves collapse into the structure you call matter.

Atoms are not physical building blocks.

They are standing scalar identities, persistent whirlpools of geometry held in place by tri-wave pressure within the Aether.

Atoms form when:

Partiki → intention sparks

Partika → intention oscillates

Particum → structure stabilises

Standing scalar waves lock into shells

These shells nest in harmonic ratios

The inner shells compress

A density threshold is reached

Matter precipitates out of Aether

Matter is not solid.

It is slowed light, folded through harmonic tension until it becomes perceivable.

This is why the truth of atoms was forbidden.

Atoms reveal that the universe is built from memory and intention, not brute mechanics.

Every atom remembers the tone from which it was born.

This is why matter responds to:

sound, geometry, thought, emotion, lineage, trauma, and love.

Matter is a field.

A field is memory.

Memory is identity.

This is the root of all creation.

E.2 - Species as Scalar Architectures

Every species, luminous, non-luminous, human, giant, dragon, and hybrid, arises from a unique scalar blueprint.

These blueprints are not drawings or written instructions. They are standing waves that encode:

memory, behaviour, instinct, geometry, and form.

A species template contains:

identity

purpose

instinct

behaviour

emotional tone

harmonic frequency

lineage memory

role in the world-grid

A rose becomes a rose because its scalar blueprint instructs the Aether to take that shape.

A wolf becomes a wolf because its field knows its geometry.

A human becomes a human because the blueprint carries the memory of that form.

Tartaria could read and write these blueprints.

Your age has never known them.

E.3 - Species of the Upper Bands (Light Beings)

These are the luminous beings above the firmament.

Your world once called them angels, but they are older, more varied, and more radiant.

The Celestials

High-band tri-wave intelligences composed of coherent light and scalar radiance rather than substance.

The Gatekeepers (Upper Orders)

Stewards of density boundaries and calibrators of Gate Star resonance.

Their bodies are structured from tri-wave lattices woven with harmonic light fields.

They do not die. They change frequency state.

E.4 - Expressions of the Mid-Band Species (Human and Harmonic Beings)

These beings have balanced tri-wave pressure, allowing them to exist as physical, semi-physical, or hybrid forms.

Current Humans

Human bodies contain:

- atomic shells
- emotional-frequency shells
- mental shells
- soul shells
- lineage memory bands

Humans are bridge beings designed to ascend up the ladder of realms as they remember their tone.

In the age of Tartaria there was no separate ordinary human civilisation on the surface.

The Tartarians were the human expression of that era, an evolved branch of the same core template, walking in greater coherence, stature, and range, but still human.

Hybrids

Hybrids form when two species scalar shells overlap and produce a new harmonic envelope.

Before the Fall, most hybrid lineages lived in outer bands or deep underworld domains.

Their stability depended on the full planetary grid remaining coherent.

After the Fall, when the layers collapsed and outer races mingled with the emptied surface population, new hybrid lines were born inside distortion.

These post-Fall hybrids, human plus invader lines, are unstable because their shells formed in a fractured field, not within the original tri-wave harmony.

Giants

Giants are frequency anchors, their bodies built around massive outer shells able to ground enormous Aether flows.

They are not mutated humans.

They are resonant beings created to stabilise the world-grid.

Their bones were crystalline, their fields enormous, their presence calming.

Dragons

Dragons were aeronautic tri-wave beings designed to regulate upper-air Aether currents.

Their bodies were harmonic engines:

internal plasma chambers

crystalline rib structures

wing membranes tuned to the sky-bands

conscious interaction with Aether flow

They were not monsters.

They were sky custodians.

E.5 - Species of the Lower Bands (Shadow and Inverted Lines)

These species exist on collapsed or inverted tri-wave patterns.

Their shells exhibit:

low-frequency resonance

high density of fear-signature

weakened geometric structure

parasitic behaviour

inability to self-sustain

The Draconian Outer Orders

Not the true cosmic dragons.

These are collapsed-field predators whose harmonic engines imploded.

They feed on the fields of others because they cannot renew their own.

The Reptilian Castes

Hybridised beings created in the outer realms after the great fragmentation.

Their shells run on low-frequency harmonic loops.

They cannot ascend without external assistance.

Inverted Tri-Wave Entities

Beings of collapse, tri-wave fields flipped inside out by distortion.

They cannot incarnate. They attach.

They were the ones who crossed over only after the inversion wave collapsed the planetary field.

Their movement into the surface world came only after Tartaria's Fall, when broken grids and amnesic populations made infiltration possible.

E.6 - Relationships Between Species (Field Compatibility)

All inter-species interaction is governed by scalar compatibility.

Every species carries a distinct field architecture. When two fields interact, the outcome is determined not by intent or belief, but by harmonic alignment.

Compatible fields enter resonance.

They stabilise one another, exchange energy efficiently, and increase systemic coherence.

This expresses as cooperation, mutual uplift, and shared equilibrium.

Incompatible fields enter interference.

Energy exchange becomes inefficient, asymmetrical, or extractive. This expresses as stress responses, predatory dynamics, avoidance, or structural failure.

These outcomes are not chosen.

They arise automatically from field interaction.

This is why giants calmed Tartarians.

Their broader, slower field structures dampened instability in human emotional and mental bands.

This is why dragons balanced the sky bands.

Their geometry interfaced directly with atmospheric and upper-grid harmonics.

This is why light-band beings uplifted surrounding systems.

Their fields radiated coherence outward wherever compatibility existed.

This is why inverse-band beings fed from denser fields.

Their architectures could not sustain coherence independently and relied on external energy intake to delay internal breakdown.

No interaction was moralised at the time.

No species was considered virtuous or corrupt.

Inter-species relationships were understood as **field mechanics**, not intention, belief, or ethics.

E.7 - Bridge to Section F

Now that you understand:

the beings of light

the beings of the mid-bands

the beings of the shadows

and why their fields behave as they do
you are ready for the next knowledge:
how the Gate Stars choose your path,
how souls ascend,
how scalar shells unwind at death,
and why some fall as shooting stars.

In Section F we turn to the gates, the stars, and the path of ascension itself.

Section F – The Gates, the Stars and the Path of Ascension

F.1 - Why You Must Understand The Stars

You have learned how beings differ by their scalar shells, their harmonic pressure, and their lineage.

Now you must learn how souls move between worlds, through stars, through gates, through the firmament itself.

This knowledge was once held by the Resonant Orders, the Gatekeepers, and the Tonal Cartographers.

It was forbidden after the Fall because it exposes the truth your age cannot yet reconcile:

Death is passage, not ending.

Ascension is physics, not belief.

Falling stars are souls, not stones.

I speak now so none may be lost through ignorance.

F.2 - The Firmament As A Living Membrane

The firmament is not:

glass, crystal, ice, metal, or a dome.

It is a frequency membrane, a living boundary woven from:

high density scalar waves

tri-wave locks

geometric nodes

luminous filaments

rotating harmonic gates

It does not separate realms by distance. It separates them by coherence.

To pass the firmament, a soul must reach a specific harmonic stability.

Not purity.

Not behaviour.

Not judgement.

Coherence.

The firmament is tuned to the frequencies of the Gate Stars. Only coherence opens the way.

F.3 - The Fixed Stars As Passage Nodes

Fixed stars are not burning gas.

They are:

harmonic tunnels

tri-wave vortices

consciousness gateways

lineage sorters

memory projectors

When a soul leaves its body, its scalar shells begin to unwind.

If the field is coherent, it rises along the harmonic band of the star from which its lineage descends.

Every human lineage has a star alignment.

Not astrology. Physics.

Your blood remembers the star that birthed your soul.

When death comes, that resonance awakens.

F.4 - THE WANDERING STARS (PLANETS)

Wandering stars, what you call planets, are mobile harmonic engines.

Their differences arise from:

their tri-wave rotations

their pressure bands

their scalar shell thickness

their field alignment with the Sun

their memory patterns

They feel closer to you because their frequencies intersect the human band more directly.

They influence mood, tides, thought flow, emotional turbulence, and energetic surges.

This is not superstition.

It is resonance physics.

F.5 - THE GUARDIAN STARS

Guardian stars stabilise the field.

They:

maintain realm boundaries

regulate the firmament membrane

hold the harmonic balance of species

support Gate Star coherence

respond to human emotion

flare during ascension events

Some people feel watched by certain stars.

They are not imagining it.

The Guardian Stars feel them back.

F.6 - The Gate Stars And Ascension

Gate Stars are ascensional apertures that open when a soul reaches harmonic alignment.

A Gate Star:

scans the field

identifies lineage

reads coherence

opens the correct passage

Not all stars are gates.

Only the highest harmonic bands are.

To pass:

the emotional shell must stabilise

the mental shell must clear

the causal shell must align the lineage band must activate

A coherent soul rises effortlessly.

An incoherent soul cannot pass.

This is the physics behind the ancient weighing of the heart, misunderstood as judgement.

It was always frequency.

F.7 - Shooting Stars And The Fallen Passages

This is the truth your world was never allowed to know.

A shooting star is a failed ascension.

It is not rock.

It is not debris.

It is not ice.

It is:

a collapsing soul shell

a field that failed to clear the firmament

a tri-wave that lost coherence

a memory body falling back into the lower bands

When the Gate Star rejects a soul because its field is unstable, the shell collapses and ignites against the firmament membrane.

This is not punishment.

It is simply physics.

A soul whose shell collapses:

does not cease to exist

does not lose its identity

does not fall forever

It returns to the lower bands, recovers, and ascends again when ready.

Witnessing a shooting star has always been sacred because it is witnessing a soul's return to the learning bands.

Not death, but reset.

F.8 - THE TRUE MECHANICS OF DEATH AND PASSAGE

When the body dies:

the physical shell loosens

the emotional shell unfolds

the memory body detaches

the field rises toward its lineage star

the Gate Star assesses coherence

the firmament opens or remains closed

If the firmament opens, the soul ascends to the lineage realm.
If the firmament remains closed, the soul returns through the astral bands.

There is no judgement.
Only resonance.

No one is denied forever.
Every soul rises when ready.

F.9 - Bridge to Section G

You now understand how souls move:

through the firmament, through Gate Stars, through the ladders of frequency that bind worlds together.

You know that death is a change of band, that coherence is the true measure, and that no soul is lost, only delayed.

But one question remains:

If passage is natural, if the stars are ready, if the Field is generous, why did the Light not return sooner?

Why did the Gate Stars dim for so long?
Why did the firmament harden?
Why did the world fall into an age when the heavens seemed silent?

To answer this, you must see not only the path of the individual soul, but the wound of the world itself.

Now I will show you why the Light delayed, and how the architecture of the Gate of Twelve shapes the fate of an entire era.

Section G – Why the Light Delayed

Astra speaks: the truth of the long silence.

You may ask why the Source Light has taken so long to return.

Why the Gate Stars remained dim.

Why the world lay century upon century in forgetting, as though the heavens had turned their face away.

The truth is older than memory, and gentler than the stories told in the dark.

Now understand.

G.1 - The Fall Was Not A Moment But A Collapse Of Many Layers

When the Gate Stars dimmed, they did not shatter like glass.

They folded.

A harmonic ladder spanning seven realms collapsed inward across three Harmonic Universes at once.

The firmament membrane twisted, the Aether grid convulsed, the Earth Heart faltered in its breath, and the human field fractured into dust like echoes of its former coherence.

You imagine a single catastrophe.

But the Fall was not a single blow.

It was a cascading resonance failure that rippled through:

  the planetary tri-wave

  the lineage bands

  the emotional shells

  the world-grid

the Soul Matrix

and even the Gate Stars that govern ascension

A wound in one layer can be stitched.

A wound in all layers cannot.

It must be grown back into alignment.

This is why the Light delayed.

G.2 - Time Cannot Heal What Frequency Cannot Hold

You live in the slowest band of the Ladder.

In the physical realm, time drips like heavy honey, painfully slow, painfully thick.

But the higher realms do not move in time.

They move in coherence.

Where you count centuries, they feel only the flicker of a breath.

When the Gate Stars fell out of phase, their time bands could no longer touch yours.

Light travels instantly in its own realm,

but cannot cross into distortion without consequence.

The long wait was not absence.

It was protection.

Had the Light descended before the membrane healed, your world would have collapsed under its radiance.

The Veil guarded you from a brilliance you were not yet ready to receive.

G.3 - The Light Cannot Enter A Field At War With Itself

The Source Light does not conquer.

It aligns.

It does not descend through force.

It descends through coherence.

And so long as humanity was fractured, so long as the emotional shells churned, the mental fields distorted, and the lineage cords lay dormant, the Gate Stars remained silent.

Not in punishment.

In physics.

A world whose tri-wave is broken cannot receive higher bands without shattering like a stone struck by thunder.

The Light delayed because it will never violate the natural order of ascent.

G.4 - The Firmament Membrane Had To Heal

The firmament is not a dome.

It is a living tension between realms, a luminous density boundary woven from harmonic pressure and standing scalar waves.

When the Fall came, the membrane folded upon itself.

This sealed the realm, not like a lock, but like a wound closing over.

To reopen it prematurely would have torn the planetary field apart.

The membrane had to soften, brighten, thicken, and regain its former symmetry.

This demanded:

the slow rise of human coherence

the gradual repair of the world-grid

the return of emotional stability

the awakening of lineage memory

The membrane did not resist the Light.

It was healing from the trauma of collapse.

G.5 - The Grid Could Not Sing Its Note

The world-grid is the Earth's instrument.

It is the resonance body of the planet, and humanity's extended nervous system.

After the Fall, the grid went silent.

It hummed only in broken intervals.

Discordant, unstable, unable to hold the song of the Aether.

The Gate Stars listen for this planetary chord before opening.

And for a very long age the Earth could not sing.

Cities rose without harmony.

Nations fractured their own lines.

Humans forgot their tone.

Children arrived with shells too thin to remember home.

The Gate Stars waited for the grid to find its breath again.

Every healing human contributed to this return.

Every awakening heart strengthened the chord.

Every moment of coherence was a note returned to the world song.

When enough tones aligned, the Light began moving again.

G.6 - Humanity Had To Choose Its Own Ascent

The Source never withdraws choice.

Ascension cannot be imposed.

A Gate Star will not reopen to a species that does not reach toward it.

The Light delayed because humanity had not yet decided between remembrance and distortion.

Every age presents a choice:

rise or sleep,

remember or remain.

For many centuries after the Fall, the human field was too dim, too frightened, too divided to rise.

But in this era, your era, the decision has shifted.

Fear loosens.

Memory stirs.

Intuition surges.

The human field begins to lift itself.

And when a species chooses ascent, the Gate Stars answer.

G.7 - The Quickening Required The Return Of Kalai-Mur

Before the Gate Stars can reignite, the Flame had to return.

And before the Flame could rise, a greater harmonic had to answer.

That harmonic is **Kalai-Mur**.

Kalai-Mur is the ancient quickening field of restoration, a vast cosmic resonance that moves through creation when a world calls not in despair, but in coherence.

It is not sent to rescue, nor does it arrive to judge.
It responds when a realm's underlying order becomes audible again to the greater field.

The Earth cried for restoration.

Not with words, but with resonance.
With pressure held too long.
With imbalance seeking release.
With a grid straining to remember its original alignment.

That cry was heard.

Kalai-Mur entered the system through the Sun.

The Sun does not originate **Kalai-Mur**.
It receives it, harmonises it, and broadcasts it into the planetary field, translating a cosmic tone into a form the Earth can receive.

Through this solar translation, the Amoraea Flame within the Earth was called to remember itself.

The Flame does not arise spontaneously within a fallen world.
It remembers only when the greater harmonic calls it into coherence.

Without **Kalai-Mur**:

the planetary grid cannot stabilise,
the dimensional membrane cannot thin,
the lineage bands cannot reconnect,
the Flame cannot rise.

This is why the Flame does not return until there is enough coherence for it can anchor.

Not because restoration is withheld, but because coherence must reach the threshold where restoration can be sustained.

Kalai-Mur does not arrive at random.
It arrives when the mathematics of the realm align with the memory of its original design.

Restoration follows law, not urgency.
It waited, as it always does, for the proper sequence to resume.

When the Gate of Twelve began to hum again beneath the soil, it marked the moment the Earth grid could once more receive the broadcast.

Only then could the Flame respond.
Only then can the Gate Stars begin to stir toward reignition.

G.8 - The Light Returned When Humanity Could Bear It

You imagine the heavens watching you in silence.

They were not silent.

They were patient.

They waited for:

enough souls to awaken

enough coherence to stabilise

enough memory to return

enough distortion to dissolve

enough hearts to rise above fear

You live now in the moment when the long delay ends.

The Gate Stars brighten.

The membrane thins.

The grid sings.

And the Light breathes again through cracks that once were fractures,

through tones that once were silent.

The delay was long.

But the return is certain.

The Light has never been late.

It arrives exactly when the world is ready to rise with it.

G.9 – Bridge to Section H

You now understand why the Light delayed, not through absence, but through law.

You have seen that restoration could not begin until coherence returned, the grid found its tone, and the greater harmonic answered the Earth's cry.

But the return of Light alone does not restore a world.
Light must enter through structure.

Before ascension can resume, before Gate Stars can fully reignite, the planetary field itself must awaken the organs designed to receive and distribute the descending tone.

These organs are the Twelve.

In Section H, I will show you the Gate of Twelve, the living nodes beneath your world, and the precise roles they play in restoring balance, memory, and ascension to a once fallen realm.

Section H – The Gate of Twelve: The Twelve Nodes And Their Purposes

Astra speaks: the hidden architecture of the ascending world.

You have learned what the Gate of Twelve is.

Now I will show you what it does.

Each node is not a point on a map, but a living organ of the planetary field, a conductor of tone, memory, and ascension.

Together they form the Great Chord, the harmonic structure through which the Source Light descends and through which souls rise in return.

Walk with me through the Twelve.

H.1 - The First Node: Growth

The Green Gate, the pulse of becoming.

This node governs the generative forces of the world:

the upwelling of life, the emergence of new forms, the pulse by which plants grow, bodies heal, and consciousness expands.

Its tone is a bright emerald harmonic, felt as a rising warmth beneath the ribs.

When active, the world grows effortlessly.

When dim, life struggles and the fields feel tired.

This node awakens first in every restoration cycle.

H.2 - The Second Node: Memory

The Golden Gate, the chamber of remembrance.

Here lies the lattice of planetary memory:

ancestral lines, soul lines, grid lines, forgotten epochs, and the records of past worlds.

Its tone is amber gold, ringing like metal warmed by sunlight.

When active, lineage memory awakens, dreams sharpen, intuition flares, and forgotten knowledge rises unbidden.

When dim, humanity forgets itself.

This node governs the return of wisdom.

H.3 - THE THIRD NODE: WATER

The Blue Gate, the keeper of flow.

This node regulates all liquids of the world:

oceans, rivers, blood, lymph, the emotional field itself.

Its tone is deep sapphire, felt in the lower belly as a gentle internal tide.

When active, water becomes a carrier of clarity, cleansing distortion across the grid.

When dim, stagnation spreads, storms become chaotic, and emotions become heavy and unclear.

This node governs purification and emotional coherence.

H.4 - THE FOURTH NODE: FIRE

The Red Gate, the pulse of vitality.

This node governs metabolic heat, volcanic breath, plasma currents, and the internal flame of consciousness.

Its tone is crimson gold, a spark behind the sternum that wakes courage and will.

When active, vitality rises through every living thing.

When dim, exhaustion spreads across cultures, and people lose the inner fire to rise.

This node governs strength and activation.

H.5 - THE FIFTH NODE: TIME

The Silver Gate, the keeper of rhythms.

This node maintains the rhythm by which events unfold:

the pacing of seasons, the cycles of incarnation, the unfolding of destiny timelines.

Its tone is silver white, a cold ringing like struck crystal.

When active, time flows smoothly, synchronicities align, paths open, and destiny feels coherent.

When dim, time becomes turbulent and compressed, feeling too fast or too slow, and human perception fractures.

This node governs the rhythm of experience.

H.6 - THE SIXTH NODE: MOVEMENT

The Indigo Gate, the path maker.

This node governs all motion:

winds, tectonics, blood flow, migration patterns, and the movements of souls between bands.

Its tone is indigo blue, felt as a subtle forward pull behind the spine.

When active, obstacles dissolve easily.

When dim, stagnation afflicts both the land and the human field.

This node governs momentum, travel, and the smooth flow of evolution.

H.7 - THE SEVENTH NODE: BALANCE

The White Gate, the great stabiliser.

This is the spine of the Gate of Twelve.
It holds the planetary tri-wave in equilibrium.

Its tone is pure pearl white, a soft pressure at the centre of the forehead that creates stillness.

When active, the world enters harmonic neutrality.
When dim, polarity extremes rise:

fear versus faith,

anger versus apathy,

chaos versus order.

This node governs neutrality, peace, and the stability of the planetary field.

H.8 - THE EIGHTH NODE: LIGHT

The Solar Gate, the first descent of radiance.

This node regulates the entry of higher frequencies from the Guardian Stars and the Source.

Its tone is radiant gold white, felt as warmth along the spine and clarity in the eyes.

When active, the atmosphere brightens, auroras intensify, and consciousness lifts.

When dim, the sky feels heavier, and humanity loses sight of the higher realms.

This node governs illumination and ascensional readiness.

H.9 - The Ninth Node: Form

The Crystal Gate, the architect of matter.

This is the node that failed first in the age of Tartaria.

It governs geometry, matter formation, the crystalline lattice of the world, and the physical shells of all beings.

Its tone is crystalline lilac, a faint humming in the bones.

When active, matter holds coherence:

cities stand stable, bodies heal quickly, the Earth remains balanced.

When dim, structures weaken, bodies age faster, crystals lose memory, and the planetary grid destabilises.

This node governs stability of form, the foundation of the physical world.

H.10 - The Tenth Node: Sound

The Harmonic Gate, the voice of the grid.

This node governs vibration itself:

language, music, communication, emotional resonance, the language of species, and the Aetheric tones that guide the dragons and giants.

Its tone is iridescent violet silver, felt as a ring in the inner ear.

When active, harmony spreads easily, and the grid speaks clearly through intuition.

When dim, miscommunication floods the world, species fall out of contact, and the Choral Exchange collapses.

This node governs harmony, unity, and the clarity of perception.

H.11 - THE ELEVENTH NODE: HEALING

The Jade Gate, the weaver of restoration.

This node governs regeneration of all kinds:

cellular, emotional, atmospheric, planetary, and Aetheric.

Its tone is green white, felt as a cool breath at the crown of the head.

When active, wounds heal rapidly, in the body, in the grid, in the collective heart.

When dim, healing slows and trauma embeds more deeply.

This node governs restitution, return to order, and the dissolving of distortion.

H.12 - THE TWELFTH NODE: SPIRIT

The Violet Gate, the crown of ascension.

This is the highest node, the bridge between the planetary field and the Source current.

Its tone is violet gold, felt as a rising expansion in the chest and a widening of perception.

When active, the firmament softens, Gate Stars open, and ascension becomes possible.

When dim, the world feels spiritually muted, souls feel cut off, and humanity forgets its origin.

This node is the final to awaken and the first to close in any Fall.

When it brightens again, the Gate of Twelve begins its return.

H.13 - WHEN THE TWELVE SING AS ONE

When all twelve nodes align:

the world-grid lights like a single instrument

the firmament becomes permeable

the Gate Stars ignite

the higher bands descend

humanity remembers its original stature

This moment is the event myth called the Opening of Heaven, Tartaria called the Great Alignment, and I call simply, the Return.

It begins with a whisper, a memory, a rising of intuition, a pull toward old lands, a softening in the field.

It ends with the entire world remembering its tone.

H.14 - THE REAWAKENING OF THE TWELVE (THE CURRENT CYCLE)

Astra speaks: the Twelve stir beneath you, and the world begins to remember.

You live in the era the silent ages prepared for.

The long sleep is ending.

The Veil thins.

The grid hums faintly again beneath your feet.

The Gate of Twelve, dormant since the Fall, begins its slow and unstoppable return.

This chapter reveals the stages of its reawakening and the signs already present in your world.

H.15 - THE FIRST STIRRING: THE RETURN OF RESONANCE

The first sign of reawakening is not light, nor prophecy, nor sudden miracles.

It is resonance.

A soft vibration rising through the soil, through the human emotional field, through dreams, through intuition, through the atmospheric band itself.

This stirring is subtle.

A whisper rather than a sound.

But once it begins, it does not cease.

It marks the moment the planetary tri-wave is strong enough to call the Twelve out of dormancy.

This began in your era.

H.16 - THE SECOND STIRRING: RETURN OF THE FLAME

The second phase is the descent of **Kalai-Mur**, the Quickening Flame.

It arrives invisible, yet unmistakable.

You know it by:

the speeding of inner growth

the intensifying of emotional clarity

the collapse of old patterns

the acceleration of self awareness

the dissolution of fear systems

the awakening of dormant memory

Kalai-Mur restores the phase lock between Earth and the higher bands.

Without it, not one of the Twelve could rise.

The Flame arrived when humanity's coherence reached threshold.

You feel its presence as an inner pressure that refuses sleep.

H.17 - The Third Stirring: Node Synchronisation Begins

Once the Flame stabilises the tri-wave, the first three nodes respond:

Growth (Green Gate)

Memory (Golden Gate)

Water (Blue Gate)

Growth rises first. You feel it as a sudden acceleration in collective evolution.

Memory rises second. Ancestral knowledge awakens, timeless dreams return, lost intuitions resurface.

Water rises third. The emotional field begins to purify, bringing long buried distortions to the surface.

Humanity is experiencing these now.

These are the first three chords of the returning Great Song.

H.18 - The Fourth Stirring: The Rise Of The Red Gate

When the first triad stabilises, the Fire Node ignites.

Its reawakening is unmistakable.

It manifests as:

global restlessness

surges of courage

uprisings against oppression

a refusal to remain asleep

an inner fire that demands truth

The people of this world are now in this stage.

The Red Gate has awakened.

H.19 - The Fifth Stirring: Time Begins To Soften

Next, the Silver Gate of Time stirs.

You feel it as:

the collapse of linear certainty

the melting of rigid timelines

strange loops of synchronicity

time speeding up or time disappearing

sudden shifts in destiny

events that compress or expand unnaturally

Time is not breaking.

It is unbuckling so that higher band pathways can rejoin the world line.

The Silver Gate is half awake.

H.20 - The Sixth Stirring: Movement Accelerates Globally

Once Time softens, the Indigo Gate of Movement begins its activation.

Signs include:

rapid migrations

sudden relocations

global unrest

tectonic shifts

atmospheric anomalies

career and life path upheavals

The world begins to rearrange itself to match the rising harmonic.

Nothing stays still.

Nothing remains how it was.

Movement is the engine that pushes humanity into new alignment.

This gate is rising now.

H.21 - The Seventh Stirring: The Rise Of Balance

Here lies the turning point.

The White Gate of Balance begins to activate.

Humanity will feel this not as peace, but as tension between extremes reaching a breaking point.

The rise of Balance always brings:

polarisation

intensified division

truth revealed forcibly

illusions collapsing

systems crumbling

personal and collective reckoning

This is not disharmony.

It is correction.

When the White Gate stabilises, polarity dissolves and neutrality returns.

You stand in the early threshold of this phase.

H.22 - The Eighth Stirring: The Atmosphere Brightens

Once Balance is regained, the Solar Gate of Light awakens.

Its signs:

brightening skies

unusual auroras

pillars of light

twin sun phenomena

changes in atmospheric density

sudden clarity in consciousness

rapid intuitive expansion

The air itself becomes luminous.

Human perception will shift upward.

The Veil will thin.

You are beginning to see this already.

H.23 - The Ninth Stirring: The Crystal Gate Recovers

This is the most crucial step because the Node of Form was the first to fall.

It must also be the one which stabilises before the higher nodes can fully awaken.

Signs of its return:

rapid healing in the body

crystalline dreams

wide scale interest in sacred geometry

resurfacing of ancient structures

discovery of hidden chambers and grids

personal physical transformation

The world's matter field begins to remember its original precision.

This node is stirring but not yet stable.

H.24 - The Tenth Stirring: Sound Returns To Purity

Once Form stabilises, the Harmonic Gate of Sound will awaken fully.

Its signs will be unmistakable:

species communicating across frequency

the return of the Choral Exchange

humans hearing tones that others cannot

sudden global interest in vibration based healing

buildings and stones resonating again

dragons and giant line souls feeling the call

The world will begin to sing again.

You are approaching this threshold.

H.25 - The Eleventh Stirring: The Healing Wave

Next, the Jade Gate of Healing rises.

Its signs:

sudden global remission events

rapid recovery processes

dissolution of deep trauma

atmospheric purification

water returning to clarity

waves of peace after turbulence

The grid begins self repair.

The human field follows.

The Healing Node awakens only when humanity has faced its shadow and refused to remain broken.

This gate is on the near horizon.

H.26 - THE TWELFTH STIRRING: THE RETURN OF SPIRIT

When the first eleven nodes stabilise, the Violet Gate of Spirit awakens.

This is the moment the Gate Stars ignite.

Its signs:

perception of higher beings, without distortion

thinning of the firmament

visible Gate Star vortices

souls regaining memory of origin

the sensation of ascending while awake

the return of the higher blueprint of humanity

This is the gate of ascension.

It is the crown of the Twelve.

Its awakening marks the end of the long silence.

The Violet Gate has not opened yet, but its shadow touches your era.

Its breath is on the horizon.

Its pulse approaches.

H.27 - The Great Alignment: When All Twelve Sing Once More

When the Twelve align fully:

the planetary resonance returns to pre Fall coherence

the firmament loosens its density

the Gate Stars brighten into visible pillars

the Guardian Stars re establish contact

the human field lifts out of inversion

the Source Light descends like a second dawn

This is the event your ancestors foresaw.

This is the cycle the Tartarians tended.

This is the return inscribed in stone, sky, and memory.

You live in its beginning.

The Gate of Twelve awakens beneath you.

Rise with it.

H.28 - The Hidden Thirteenth Gate

Astra speaks: The secret held beneath the Twelve.

There is a truth beneath the Gate of Twelve that few worlds ever learn, and fewer still survive long enough to remember.

The Twelve are not the whole.

They were never the whole.

Beneath their circle lies the Central Flame, the Silent Axis, the Gate Behind the Gates, the Thirteenth Gate.

A gate with no direction, no colour, no node, no element, no star.

A gate that does not open outward into realms above, but inward into the Origin.

Now you will learn what even the First Humanity only partly understood.

H.29 - THE THIRTEENTH GATE WAS NEVER MEANT TO OPEN DURING AN AGE

The Twelve govern the realm.
The Thirteenth governs the root of the realm.
The Twelve shape:
life,
form,
memory,
movement,
healing,
ascension.
But the Thirteenth shapes:
the existence of the realm itself,
the bond between Source and world,
the continuity of the planetary soul,
the stability of the Great Ladder,
the very persistence of time and matter.
It is the Gate of Gates, the axis on which the world is hung.
The Twelve are petals, the Thirteenth is the stem.
The Twelve are spokes, the Thirteenth is the hub.
The Twelve awaken for ascension cycles.
The Thirteenth awakens only for world cycles.

It should not open in time.

It should only open at the beginning and end of entire aeons.

But in your age, it began to stir.

And this is why the Fall was catastrophic.

H.30 - THE THIRTEENTH GATE IS THE HEART OF THE HAAR

The HAAR - the Heart of the Aetheric-Auric Resonance - is not merely a planetary organ.

It is the outer expression of the Thirteenth Gate.

Think of the HAAR as the visible glow of an invisible source.

What you know as:

the Earth Heart,

the Deep Pulse,

the resonance core,

the tone beneath all tones

is simply the surface shimmer of the Thirteenth Gate pressing upward into matter.

The Thirteenth Gate is the Source-thread anchored into your world.

When the HAAR faltered, it was not merely failing.

It was protecting the Thirteen from tearing open in instability.

This is why giants descended into the deep.

This is why the Mirror Veins hummed.

This is why the grid screamed in the final hours before the Fall.

The Thirteenth Gate was beginning to open, and it must never open in distortion.

So the world collapsed inward to seal it.

This was survival, not punishment.

The HAAR is the central harmonic interface where the aetheric structure of reality and the auric fields of living beings converge.

It is not atmospheric alone, nor biological, nor abstract.

It is the living resonance core through which worlds stabilise, consciousness incarnates, and memory is maintained across realms.

H.31 - The Thirteenth Gate Is the Only Gate That Faces Two Ways

Every GateStar, every node of the Twelve, aligns upward into higher bands.

But the Thirteenth Gate is the only one that aligns:

upward to the Source,

downward to the planetary soul,

inward to the Origin,

outward to all timelines simultaneously.

It is the axis of continuity.

If it opens correctly, a world ascends as a whole into its next configuration.

If it opens incorrectly, a world collapses into its own shadow and becomes the seed for a future reality.

Your world survived because the Thirteenth did not open fully.

But its partial awakening was enough to tear the firmament, invert the emotional shell, and collapse the entire GateStar network.

H.32 - What Lies Beyond the Thirteenth Gate

Humanity has legends of a place beyond light, beyond the stars, beyond birth.

All these myths speak of the same truth:

Beyond the Thirteenth Gate lies the Origin Field.

Not "heaven,"

Not "source,"

Not "other realms."

The Origin Field is the point from which all timelines arise and converge.

It is:

the root of all Gate Stars,

the blueprint of all ladders,

the seed of all universes,

the silent well from which all souls emerge,

the field where no inversion exists,

the memory before memory.

Worlds rarely touch it.

When they do, their entire existence reshapes.

This is why the Thirteenth Gate remains hidden.

It is too powerful to open during ordinary ages.

H.33 - Why the Thirteenth Stirred During the Fall

The Fall of Tartaria was not the failure of a single node, nor merely the collapse of a grid, nor only an atmospheric or energetic inversion.

It was a convergence of failures so rare that it forced the deepest safeguard of the world to respond.

There were three causes, woven together in a sequence that had never occurred on this world before.
I will reveal them plainly.

Cause One - The Collapse of the Ninth Node (Form)

When the Ninth Node, the Node of Form, drifted out of alignment by a fraction of a tone, the geometric framework of the world began to buckle.

This was not immediately visible, but its effects cascaded rapidly.

The towers misaligned.
The grid destabilised.
The HAAR pulse faltered.
The firmament twisted.
Matter itself began to vibrate out of balance.

A structural vacuum formed within the outer ring of the world-grid.

The Thirteenth Gate sensed this failure immediately.

This is not intelligence in the human sense, but a protective reflex embedded in every complete world-grid. When structural continuity is threatened, the Origin-thread responds.

This was the first stirring of the Thirteenth Gate.

Cause Two - The Collapse of the Emotional Shell

In the same window of collapse, the Water Node fell out of harmonic balance.

When a world's emotional field collapses, the effect does not remain at the surface. It sends a resonance shock downward into the deepest layers of the grid.

This shock-wave travelled through the Mirror Veins and into the lower chambers of the HAAR.

The Thirteenth Gate felt this distortion as internal turbulence. The Gate shuddered inwardly and its activation deepened.

Had this emotional collapse occurred after the failure of Form, the Thirteenth Gate would not have stirred.

But the collapse of structure and the collapse of emotional coherence happened simultaneously.

This has occurred only three times across all recorded timelines.

Cause Three - The Twisting of the Firmament Membrane

This was the decisive moment.

When the firmament membrane twisted out of phase, the planetary field lost its upward tension. The world sagged inward, energetically speaking, and the weight of the grid pressed toward the centre.

Resonance was forced into the HAAR faster than it could release.

The Thirteenth Gate began to take on the excess load.

This was not an opening driven by expansion. It was a fail-safe response.

The partial opening of the Thirteenth Gate was an attempt to prevent total planetary implosion.

It succeeded, but barely.

The Consequence of the Partial Opening

The Thirteenth Gate absorbed the shock of three simultaneous failures.

But it could not open fully without unmaking the realm itself.

So it opened only enough to stabilise the core and seal the collapse.

The price was severe.

The Twelve collapsed.
Memory fractured and withdrew.
The emotional field inverted.
The Gate Stars dimmed.
The long silence you call the ages after began.

The Thirteenth Gate did not save Tartaria.

It saved the world.

The Fall was not punishment, nor error, nor hubris.

It was the lesser destruction chosen to prevent the greater.

H.34 - The Thirteenth Gate Is Stirring Again - But Correctly This Time

During the current cycle, the Twelve are reawakening in the correct order.

This time, the Thirteen does not stir in warning.
It stirs in welcoming.
You feel this as:
a sense of destiny accelerating,
an internal pull toward something vast,
the sensation of "being remembered,"

moments where reality feels thinner,

the rising unity of soul-lines,

the deep emotional purging of humanity,

the increase in global synchronicity.

This era is not the Fall repeating.

It is the Fall reversing.

The Thirteenth Gate awakens now because the Twelve are returning to alignment.

It awakens not in fear but in resonance.

H.35 - When the Thirteenth Gate Fully Awakes

When the Thirteen opens through harmony rather than collapse, your world will:

rejoin the higher band continuum,

regain pre-Fall perception,

experience visible GateStar activation,

enter multi-band consciousness,

reclaim memory across lifetimes,

achieve field-coherence with its own Source-thread,

and ascend as a unified harmonic body.

This is the great transition that Tartaria prepared for but never reached.

You live in the age where this possibility returns.

The Thirteenth Gate does not open first.

It opens last, as the crown of the Twelve and the anchor of the next world.

H.36 - The Thirteenth Gate Is Not a Place - It Is a Revelation

No pilgrim may walk to it.
No map may chart it.
No stone can house it.

The Thirteenth Gate opens in one location only:
the centre of the human field when the Twelve are awakened within it.

For the planetary gate is only the external reflection of the internal one.

When humanity remembers its tone, the world remembers its path.
When the world remembers its path, the Thirteen opens.
When the Thirteen opens, the era changes forever.

Thus I tell you:

The Thirteenth Gate is the moment a world realises it is no longer asleep.

You approach that moment.

H.37 – Bridge to Section I

You have now seen the Gate of Twelve as it truly is: not myth, not map, but living planetary anatomy.

Each Node is an organ of coherence, and together they form the Great Chord by which the world receives higher tone and returns its own.

But the Nodes do not awaken in isolation.

Their reactivation changes the boundary above you, because the firmament is the interface that responds to their collective song.

As the Twelve rise into synchronisation and the Thirteenth Gate stirs in harmony rather than fear, the sky itself begins to transform. The hardened membrane you were born beneath starts to soften, thicken, brighten, and return toward its original permeability.

In Section I, I will show you the New Firmament, the returning sky, the signs of its change, and what humanity will witness as the heavens begin to remember their true structure.

Section I – The New Firmament

(The Returning Sky)

Astra speaks: The sky you were born beneath is not the sky that will return.

The firmament you know today is not the original sky.
It is the bruised echo of one.

A hardened membrane, a sealed boundary, a protective shell formed during the Collapse to prevent the Thirteenth Gate from tearing fully open.

This is why the world feels "closed," why the heavens seem distant, why the Gate Stars vanished, why the higher bands are unseen.

But now, as the Twelve Nodes reawaken and the Thirteenth Gate stirs in harmony rather than fear, the firmament begins its transformation into the New Sky.

I will tell you what is to come.

I.1 - The Firmament Is a Living Boundary, Not a Dome

In the ancient age,
the firmament was:

soft,

luminous,

permeable,

resonant,

responsive.

It breathed with the world.

It shimmered with colour.

It carried the tone of the Twelve Nodes across continents.

It was not a barrier but a membrane of harmony, the interface between physical perception and the higher fields.

When the Fall came, this membrane hardened like water turning instantly to ice.

It did not shatter.

It thickened.

This was the world's instinctive defence against the Thirteenth Gate tearing open through distortion.

But the sky you see today is not the true sky of your lineage. The true sky is preparing to return.

I.2 - The Firmament Softens in Response to Coherence

As the Twelve Nodes reactivate, the membrane begins to soften.

This is not metaphor.

It is physics.

A softened firmament:

lets more Aetheric light through,

responds to harmonic pressure,

becomes semi-luminous,

reveals upper-band structures,

allows GateStar ignition,

restores multi-layer visibility.

You have already seen the first signs:

strange brightness in the sky,

more vivid colour bands at dawn and dusk,

"two suns" or mirrored star phenomena,

lights in the high air that do not behave as aircraft,

soft auroral veils appearing outside polar regions,

unusual atmospheric "holes,"

sudden clarity after storms.

These are not mysteries.

They are the first cracks in the hardened membrane.

The New Firmament is beginning to breathe.

I.3 - The Membrane Must Thicken Before It Thins

This is the paradox of the return.

Before the firmament can become soft again, it must temporarily thicken to absorb the instability released by the awakening Nodes.

During this phase, humans experience:

pressure in the head,

ringing in the ears,

sudden emotional surges,

vivid dreams,

fatigue waves,

expansions of intuition,

distortions in time perception,

moments of heightened clarity,

moments of strange inner silence.

This is not malfunction.

It is recalibration.

The membrane thickens to protect you while the Twelve rise. Then, as the Nodes synchronise, the membrane thins to reconnect you with the higher bands.

The thickening is temporary.

The thinning is permanent.

I.4 - THE FIRST VISIBLE CHANGE: ATMOSPHERIC LUMINOSITY

As the New Firmament begins to form, the atmosphere will take on a subtle internal glow.

You will see:

light that seems to come from within the air itself,

a softening of shadows,

halos around the sun and moon,

iridescent clouds,

colours that do not exist in the old spectrum,

vertical shafts of light at dusk,

clouds that glow at night without moonlight,

atmospheric "sheens" or gradients.

This is the Light Node synchronising with the firmament's returning permeability.

The sky will look "alive."

Because it is.

I.5 - The Second Change: GateStar Reappearance

When the membrane reaches stability, the Gate Stars will begin to show faintly through the upper bands.

They will appear as:

stationary points of light,

structured clusters,

slow-turning spirals,

rotating "lanterns,"

orbs that shimmer but do not fall,

geometric asterisms not found in modern star charts.

These are not spacecraft.

They are the stellar guardians that maintain the world's harmonic envelope.

In the ancient age, they were visible nightly.

They are returning.

I.6 - The Third Change: The Sky-Bands Return

Above the clouds lie the seven atmospheric bands, layers of Aetheric colour that once glowed like soft auroras night and day.

These bands include:

the Rose Band (emotional field),

the Blue Band (Aether flow),

the Gold Band (memory field),

the Silver Band (time field),

the Green Band (life-grid),

the Violet Band (spiritual access),

the White Band (planetary balance).

These bands were visible in Tartaria as shimmering veils layered above the world.

When they return:

the sky will seem "layered,"

colours will move fluidly,

the air will hum faintly,

the horizon will show spectral currents,

and the world will appear multidimensional.

This is the New Sky.

I.7 - The Fourth Change: The New Firmament Becomes Permeable

When the firmament reaches full coherence, the membrane no longer seals the realm.

It interacts.

You will see:

projected light from the upper realms,

short-lived portals that look like oval distortions,

lines or grids of light "behind" the sky,

descending bands of tone (visible as shimmer),

upward arcs of plasma that vanish mid-air,

"windows" that flicker briefly then close,

soft geometric scintillation, especially at sunrise.

This is the moment the world is ready for GateStar activation.

The New Firmament is no longer a wall but a bridge.

I.8 - The Fifth Change: The First Opening of the Gate Stars

When the New Firmament reaches its final structure, the Gate Stars ignite.

You will see:

pillars of vertical light rising from the horizon,

multi-coloured spirals in the upper air,

concentric rings spreading across the sky,

star-like points that pulse rhythmically,

luminous vortices,

the return of the "heavenly roads" of old,

the reappearance of the higher architecture that once guided dragons and sky-craft.

This is the moment of transition between the old world and the next.

The sky returns first.

Then the world follows.

I.9 - Why the Sky Must Change Before the World Changes

Because the firmament is the interface between realms.

If the sky remained sealed, no amount of awakening, no grid reactivation, no Node alignment could connect the world to the higher continuum.

The sky changes so the world can remember.

The sky reopens so the Gate Stars can return.

The sky breathes so the Thirteenth Gate can awaken fully without repeating the Fall.

The New Firmament is the promise that the Collapse will not happen again.

I.10 - How You Will Feel When the Firmament Fully Returns

You will feel:

lighter in the body,

clearer in thought,

more present,

more connected,

more intuitive,

more aware,

more coherent,

more alive.

Your dreams will sharpen.

Your senses will expand.

Your perception will widen.

Your memory will return.

You will feel the world breathing with you and you with it.

This is the rising of the New Sky.

This is the return of the ancient world.

This is the sign that the Twelve stand ready and the Thirteenth is awakening in peace.

I.11 - The First Opening Of The Gate Stars (What Humanity Will Witness)

Astra speaks: The moment when the world remembers the heavens.

The return of the Gate Stars is not a myth, not a metaphor, not an allegory.

It is a celestial event.

A planetary event.

A harmonic event.

It is the moment a world is recognised by the continuum again.

Below is what humanity will witness when the Gate Stars open for the first time since the Fall.

I.12 - The Light at the Horizon

The first sign does not appear overhead but at the horizon.

A vertical beam of pale light will rise from a point where no sun stands.

It will shimmer softly, like a luminous column made of mist.

This light is not physical.

It is the first projection of the GateStar lattice through the New Firmament.

At first you will think it is a sun-dog, or an atmospheric reflection.

But then more will appear.

I.13 - The Second Sign: The Symmetry

A second beam will rise directly opposite the first.

Then a third, a fourth, a fifth.

Each one equidistant, perfect in placement, matching the geometry of the Twelve Nodes.

Humanity has not seen this symmetry since the age of Tartaria.

The sky will form a wheel and the world will appear as though standing in the centre of a cosmic compass.

This symmetry signals that the Twelve Nodes have completed their harmonic cycle.

The Gate Stars are preparing to pierce the Veil.

I.14 - The Third Sign: The Descent of Colour

Next, colour will descend.

Not aurora.

Aurora dances erratically.

This will flow, bend, ripple, and fall like veils of liquid silk.

Seven colours, corresponding to the Seven Bands of the upper atmosphere:

Rose (emotional coherence),

Blue (Aetheric flow),

Gold (memory-field),

Silver (time-field),

Green (life-grid),

Violet (spiritual access),

White (planetary balance).

These colours will appear not at random, but in measured frequencies.

People will instinctively feel which bands resonate with them.

Some will cry without knowing why.

Some will remember.

Some will awaken instantly.

The colour-veils mark the firmament's full permeability.

1.15 - The Fourth Sign: The Rotating Lanterns

Then the Gate Stars themselves will ignite.

They will appear as slow-turning lanterns high above the atmosphere:

white-gold in tone,

crystalline in outline,

rotating with measured rhythm,

sometimes showing geometric patterns within.

These are not astral craft.

They are the stellar guardians whose presence maintains the world's link to the continuum.

Only when the Gate Stars return can ascension occur for a species as a whole.

This is the return of the lanterns of the ancient sky.

1.16 - The Fifth Sign: The Harmonic Pulse

Next, the world will hum.

You will feel it first in your bones.

Then in your chest.

Then in the air around you.

This hum is the GateStar Synchronisation Pulse, the tone that aligns the membrane with the higher realms.

It will last between 3 minutes and 9 minutes depending on alignment.

During this time:

birds will fall silent,

oceans will still,

animals will lift their heads in awareness,

the atmosphere will feel "held,"

and humanity will feel a pressure of recognition as though a forgotten presence has placed its hand upon the world.

This is the moment of re-binding between the planet and the continuum.

1.17 - The Sixth Sign: The Opening

When the harmonic pulse completes, the Gate Stars act in unison.

A circular field of light will open in the high sky.

It will appear:

translucent,

geometric,

shimmering,

soft-edged,

and vast.

Not like a portal that "sucks" or distorts.

More like a thin window seen through rippling water.

Through this field you will see higher-band structures:

crystalline ribs or arcs,

flowing currents of white-gold Aether,

distant architecture that is not physical,

light forms moving with intelligence,

sky-paths once used by dragons and craft.

This is the first opening, the rebirth of the GateStar system.

It will not stay open long, but once it has opened once, it will open again.

The new sky is a sky with doors.

I.18 - The Seventh Sign: Human Resonance Activation

Finally, the opening of the Gate Stars will trigger activation within humanity.

People will experience:

sudden clarity,

heightened intuition,

flashes of memory not from this life,

a sense of "rightness,"

a dissolving of long-standing fear,

the return of pathways within the emotional field,

the lifting of the old density shell,

the feeling: "Something ancient has returned."

This is not enlightenment.

It is recognition.

The world will know itself again.

I.19 - THE FIRST CONTACT (HOW UPPER-BAND BEINGS REAPPEAR)

Astra speaks: The return of those who once walked beside you.

The opening of the Gate Stars is not the end of the story.

It is the beginning.

When the New Firmament stabilizes and the Gate Stars ignite, the higher realms do not rush inward.

They do not descend in spectacle.

They do not overwhelm the world.

Contact happens in seven stages, slowly, deliberately, with precision and care.

Because a world waking from amnesia must be approached as gently as a child waking from sleep.

Let me show you how First Contact unfolds in an ascending realm.

I.20 - STAGE ONE: PRESENCE AT THE EDGE OF PERCEPTION

Before beings appear, their fields return first.

You will feel:

the sense of being accompanied,

a subtle warmth at your back,

a pressure in the air when you are still,

fleeting silhouettes in peripheral vision,

a rise in intuition that is not your own,

guidance that feels "placed" rather than imagined.

These are not ghosts.

They are upper-band beings whose resonance is brushing your membrane through the thinning firmament.

The world senses them before it sees them.

I.21 - Stage Two: The Return of the Harmonic Echo

As the Gate Stars strengthen, higher-band beings begin transmitting through the harmonic layer.

This does not appear as speech but as:

internal tones,

pulses of understanding,

sudden realisation,

meaning without words,

emotional clarity arriving from nowhere,

memory rising unbidden.

This harmonic echo is the natural speech-pattern of upper realms.

It is the return of the Choral Exchange.

Only those with stable fields perceive it clearly at first.

But eventually the entire species adapts.

1.22 - Stage Three: The First Shapes in the Upper Air

When the firmament reaches its permeable state, beings of higher bands can project partial forms into the visible layer.

These projections appear as:

silhouettes made of light,

luminous outlines,

softly defined humanoid forms,

winged figures without feathers,

tall shapes with radiant edges,

orbs that pulse in symmetrical rhythm.

They are not physical bodies.

They are resonance-forms cast downward through the membrane.

In ancient times, your ancestors called them:

lantern-walkers,

skycompanions,

the radiant folk,

the tall ones,

the white-shapers.

They will return in this way first, visible but not touchable.

This ensures humanity does not react in fear.

Contact adapts to your comfort.

I.23 - Stage Four: The Descent Into the Lower Atmosphere

Next comes the phase where upper-band beings can descend through the firmament
but not yet stand fully within matter.

In this stage, they appear like:

luminous figures with soft detail,

tall forms (8–14 feet) with gentle radiance,

beings whose faces seem carved from light,

semi-transparent silhouettes that move with intention.

You will perceive:

serenity rather than shock,

familiarity rather than strangeness,

calm rather than fear.

This is because your lineage has seen them before.

These beings include:

the Aetheric Guardians,

the Sky-Architects,

the First Choir,

and others who once cooperated with the Tartarian grid.

They return only when the twelve nodes are stable enough to maintain their presence without distortion.

This stage marks the return of peaceful contact.

I.24 - Stage Five: The Reappearance of Elder Species

As the resonance strengthens, the giants and dragons, the elder species aligned to the planetary grid, begin their own return.

Not all at once.
Not everywhere.
But in specific regions where the Nodes resonate strongest.

Giants return first as harmonic forms, then as semi-physical manifestations near deep resonance pillars.

Dragons return through the upper atmospheric currents once the sky-bands reopen.

Their appearance will not be chaotic.
It will be deliberate, measured, timed to coincide with the grid's stability.

Their role is not dominance.
It is guidance.

They return only when the world is ready to receive them without myth or fear.

I.25 - Stage Six: Physical Cross-Band Coherence

When the firmament reaches full alignment with the Twelve and the Thirteenth Gate,
upper-band beings can enter the physical realm in full coherence.

This does not mean they become identical to humans.
It means their bodies adjust to your density while retaining higher resonance.

In this phase, you will see:

beings tall and radiant,

with calm expressions and elongated proportion,

skin that reflects light subtly,

eyes that hold depth,

movement that seems too smooth for physical matter,

presence that stabilises the field around them.

They will be tangible.

They will speak.

They will walk.

They will teach.

And they will do so without disrupting your world's choices.

Contact becomes cooperation.

I.26 - Stage Seven: The Return of Collective Communion

The final stage is not the appearance of beings but the restoration of the Choral Field, the unified resonance between humanity and the continuum.

When this occurs:

communication is effortless,

wisdom flows both ways,

memory bridges open,

the emotional field is stabilised,

humanity reclaims its higher purpose,

the Twelve stand at full alignment,

the Thirteenth Gate awakens fully,

and the world enters

its next epoch of existence.

This is the true meaning of First Contact.

Not arrival.

Not visitation.

Not revelation.

Reunion.

A returning.

A remembering.

The beings who will come are not strangers.

They are those who walked beside you in the age before the Fall and will walk beside you again.

I.27 - After the First Opening

The Gate Stars will open intermittently at first, then cyclically, then permanently during the Great Alignment.

The return of the Gate Stars is the sign that the Twelve are complete and the Thirteenth is opening in harmony.

The world that witnesses this is already ascending.

You will live to see these signs.

I.28 - THE HOURS BETWEEN THE WORLDS (THE TRANSITIONAL AGE)

Astra speaks: The span between first light and full remembrance.

Between the first opening of the Gate Stars and the Great Alignment lies a narrow band of time.

An interregnum.

An in-between.

The hours between the worlds.

In this age, nothing is fully as it was, yet nothing is fully as it will be.

I will show you what unfolds in this passage, that you may recognise it and not mistake it for failure.

I.29 - The World That Has Seen, But Not Yet Become

After the first opening, humanity will have witnessed the sky remember itself.

But daily life will still appear, on the surface, much the same:

cities functioning,

systems struggling,

habits persisting,

old structures clinging to their forms.

This dissonance is not error.

It is the lag between inner alignment and outer manifestation. Consciousness turns faster than matter.

A world that has glimpsed the higher bands cannot return to true sleep,

but it can attempt to pretend.

For a short while.

I.30 - Intensification of the Choice

In this transitional age, the core choice of every age becomes unavoidable:

remembrance or distortion,

coherence or fragmentation,

ascent or inversion.

The opening of the Gate Stars does not remove choice. It amplifies it.

Those who lean toward coherence will find support rising beneath their feet:

synchronicities, protection, inexplicable help, sudden inner guidance.

Those who cling to distortion will feel their own patterns magnified until they break.

The Light does not punish.

It reveals.

I.31 - The Unmasking of Structures

As the planetary field re-phases with the higher bands, structures built against harmony will begin to strain.

You will see:

systems that cannot adapt collapsing rapidly,

hidden motives rising to the surface,

falsified histories cracking under their own weight,

institutions built on fear losing their binding power.

This is not the world "falling apart."

It is the world ceasing to sustain what does not match its returning tone.

In the Hours Between, unmasking is mercy.

I.32 - The Inner Firmament

While the outer sky transforms, an inner firmament will stir within the human field.

You will feel:

sudden surges of compassion, followed by grief,

waves of memory without clear image, only feeling,

phases of extreme sensitivity,

the sense that your chest has "more space" inside it,

moments of silence so deep they feel like being held.

This is your personal membrane softening.

The same physics that returns the New Firmament returns the inner sky of the heart.

In this time it is vital to:

rest often,

ground into the body,

drink clear water,

speak truth gently but firmly,

allow old pain to surface and leave.

I.33 - Early Choral Contact

Before the Choral Exchange returns in full, there will be scattered notes of its song.

Individuals and small groups will:

hear tones with no apparent source,

dream of voices singing in unison,

feel called to certain hills, coasts, towers, or stones,

begin to speak words or sounds that stabilise their own field.

These are rehearsals.

The world is re-learning how to listen.

The dragons, giants, and Choir-line souls will feel this most strongly at first.

Do not dismiss these experiences as "imagination."

Imagination is simply the first organ to perceive the returning song.

I.34 - The First Threads of the Memory-Tide

As the Nodes stabilise deeper into alignment, a tide of memory will begin to move through humanity.

It will not arrive as one clear picture but as:

persistent themes in dreams,

collective fascinations with lost ages,

sudden resurgences of ancient symbols,

children speaking of "before" with startling confidence,

discoveries in stone, ice, and archive that confirm inner knowing.

The Hours Between are the time when the First Age begins to surface through the Second, preparing you for the full remembrance to come.

I.35 - Living As a Bridge

In this transitional age, those who feel the pull of the Gate Stars most strongly are not here to escape the world.

You are here to bridge it.

To:

stand with one foot in the old, one in the new,

hold coherence while systems shake,

embody kindness in fields of fear,

remember that panic is a symptom of misread change.

You are the living fibres through which the New Sky anchors into daily life.

When you breathe, the world steadies.

When you choose truth, the grid brightens.

When you refuse to abandon your heart, the Thirteenth Gate calms.

The Hours Between the Worlds are not a delay.

They are the time granted for humanity to choose how it will arrive at the Great Alignment.

I.36 - Bridge to Section J

You have now seen what returns first in any restoration cycle: the sky.

The firmament softens, the bands reappear, the Gate Stars begin to show, and the boundary between realms becomes a bridge once more.

But a returning sky does more than change what you witness above.

It changes what rises within.

As the membrane becomes permeable, memory begins to move again through the human field, not as theory, but as recognition.
The same Gate Stars that will soon be seen also carried instruction in the First Age.
The same sky-paths that will reopen once guided dragons, towers, and the living cities of Tartaria.

To understand what is returning, you must understand what was once known.
Not as legend, but as lived civilisation.

In Section J, I will reveal the memory of the First Age, what Tartaria learned from the Gate Stars, and how that knowledge shaped a world before the long forgetting.

Section J – The Memory of the First Age (What Tartaria Learned from the Stars)

Astra speaks: the history lost beneath the Flood of forgetting.

To understand the ascent of your world you must understand the First Age, the era when Tartaria learned directly from the Gate Stars and the beings who tended them.

This chapter reveals what the First Humanity gained from the heavens before the Fall.

J.1 – The First Age Began With the Opening of the Gate Stars

In the Pre-Fall era, the Gate Stars were open nightly.

Their light:

guided the dragons,

stabilised the sky-bands,

informed the grid-architects,

governed the firmament's permeability.

To the Tartarians, the Gate Stars were not deities. They were teachers.

Their light communicated through resonance rather than language.

A pillar of GateStar light was the highest form of instruction.

J.2 – The First Lesson: Harmonic Architecture

From the Gate Stars came the knowledge of form.

This knowledge taught Tartaria:

how to tune stone so it "sang,"

how to raise towers that breathed,

how to anchor Aether currents into crystalline foundations,

how to shape geometry that stabilised the world-grid,

how to build cities that needed no fuel,

how to imbue structure with life-like resonance.

Every city was a musical instrument.

Every wall was a tuning fork.

Every plaza was a resonance chamber.

This was the architecture of the First Age.

J.3 – The Second Lesson: The Aether Sciences

From the Gate Stars came the law of Aetheric exchange.

This revealed:

how Aether condenses as light,

how sound directs energy,

how the upper atmosphere contains multi-band currents,

how levitation is achieved through harmonic counterweight,

how healing is performed through resonance alignment,

how communication travels through tone and geometry.

This knowledge birthed:

sky-craft,

canal-light systems,

atmospheric stabilisers,

resonance halls,

crystal libraries,

undersea cities.

Everything functioned without extraction or combustion.

Because Aether was the breath of the world.

J.4 – The Third Lesson: The Nine Races of Cooperation

The First Age was not human alone.

Nine cooperating species participated in the grid:

Tartarians,

Giants,

Dragons,

The Sky-Architects,

The Deep Resonants,

The Mirror Vein Striders,

The White-Band Choir,

The GateStar Attendants,

The Lattice-Born.

Each race embodied a different harmonic function.

Humans were the memory race.

Giants were the stabilisers.

Dragons were the navigators of flow.

The Choir was the tuners of the upper air.

And the GateStar Attendants were the guardians of the firmament.

From the stars came unity of purpose.

J.5 – The Fourth Lesson: The Twelve and the Thirteen

From the Gate Stars came the Revelation of the Nodes.

Humanity learned:

the Twelve govern the world,

the Thirteenth governs existence,

no species ascends alone,

no world ascends by force,

harmony is the only bridge.

The First Age was built on this understanding.

This is why Tartaria thrived without domination or hierarchy.

They understood their place in the harmonic whole.

J.6 – The Fifth Lesson: The Sky-Paths

From the Gate Stars came the knowledge of travel.

Not physical travel, but band travel:

moving between atmospheric layers,

navigating the seven bands,

entering the GateStar corridors,

descending into lower-density caverns,

ascending into the upper continuum.

Dragons mastered these paths.

Humans learned to follow.

Sky-craft became an art.

The sky was not "up."

It was layered, and each layer was a world.

J.7 – The Sixth Lesson: The Great Alignment Cycles

The Gate Stars taught Tartaria the cycles of ascent.

They learned:

when the Nodes rise,

when the firmament softens,

when the Flame awakens,

when the Thirteen stirs,

when a world enters its next epoch.

They prepared for this moment long before it came.

But the Fall interrupted the cycle.

What Tartaria learned is now returning to you.

This is the memory the world is beginning to reclaim.

J.8 – Why the Memory Returns Now

Because the Nodes rise again.

Because the Flame has returned.

Because the firmament softens.

Because the Gate Stars prepare to ignite.

Because the Thirteen awakens in harmony.

The First Age's knowledge belongs to the returning age.

History does not repeat.

Memory does.

And the memory of the First Age is rising through humanity like dawn through mist.

It returns because the world is ready for it.

J.9 – Bridge to Section K

Now you understand what Tartaria learned from the stars in the First Age:

How the Gate Stars instructed by resonance.
How architecture, Aether science, and cooperation were shaped by that guidance.
How the Twelve and the Thirteen were known as law, not belief.
How the sky was read as layered worlds, not empty distance.

But there is a missing mechanism your age does not recognise.

The Gate Stars did not teach through spoken language.
They did not send sentences.
They sent tone.

Instruction travelled through a living network of frequency, chord, and field response, shared between species, structures, Nodes, and sky-bands. The Tartarians called this network the Exchange, because it was not worship and it was not art.

It was communication.

If you do not understand the Choral Exchange, you will misunderstand the First Age itself, because nearly all instruction, navigation, calibration, and inter-species coordination occurred through it.

Now I will show you the language beneath the civilisation.

Section K – The Choral Exchange

The Language of Tone Between Worlds

K.1 – What The Choral Exchange Is

The Choral Exchange is the original language of the world.

Not words.

Not symbols.

Tone.

It is the living network of harmonised sound and frequency through which:

- species communicated across bands
- the Nodes of the Twelve synchronised
- dragons navigated the sky-bands
- giants stabilised the grid through voice and presence
- humans received instruction from the higher realms

In the First Age, every realm spoke through it.

It was not a song for entertainment.

It was the planetary nervous system in sound.

Where your age uses:

wires, signals, cables, satellites, transmissions, speech and written code,

Tartaria used:

tone, chord, resonance, and living harmony.

This was the Choral Exchange.

K.2 – How The Choral Exchange Worked In The First Age

In the age before the Fall, the Choral Exchange functioned like a living orchestra.

Three primary layers interacted:

The Planetary Layer

The Gate of Twelve, the HAAR, the firmament, and the grid-lines held a constant background chord.

The Species Layer

Dragons, giants, humans, Choir-line beings, aquatic races, and hidden lineages each contributed their harmonics.

The Stellar Layer

Gate Stars and Guardian Stars sent guiding tones through the firmament into the planetary song.

Communication travelled not through distance or volume, but through field coherence and resonance alignment.

A giant could hum a note on one continent and a dragon riding the upper bands on another continent would feel it as a navigational signal.

A Choir hall in Tartaria could sound a chord and:

towers adjusted their field output

canals altered their Aetheric flow

sky craft corrected their alignment

healing chambers recalibrated to the new tone

One song could retune an entire city.

One chord could stabilise a storm.

One unified chorus could shift the emotional field of a whole region.

K.3 – The Anatomy Of A Chord

Choral communication is layered.

Every chord contains:

    A Base Tone

The planetary or local Node frequency.

Example: Green Gate growth, Blue Gate water, White Gate balance.

    A Lineage Tone

The specific harmonic of the species or soul-line speaking.

    An Intent Tone

The emotional and mental quality carried in the sound.

    A Modulation Pattern

The rhythm and movement of the tone, which encodes meaning.

To hear only the base tone is to hear a note.

To hear all layers together is to hear a message.

    In the First Age, humans trained to:

    sense base tones in the land and sky

    recognise lineage tones of different beings

    feel intent tones in the body

    read modulation patterns as clear instruction

    They did not "translate" in the way you translate languages.

They attuned.

K.4 – The Roles Of The Species Within The Choir

The Choral Exchange was never human alone.

Each participating race held a specific musical function.

Humans - The Rememberers

They held the memory-tones.

Their voices carried story and lineage.

They encoded wisdom in song-lines, chants, and spoken harmonics.

Giants - The Grounding Bass

Their deep tones stabilised:

- tectonic plates
- grid nodes
- tower networks
- lower atmospheric bands

When giants gathered to sound together, entire regions calmed.

Dragons - The Sky Melody

Their voices rode the upper air.

With their wing membranes and internal plasma chambers, they:

- tuned the sky-bands
- cleared Aetheric storms
- opened and closed pathways in the upper atmosphere

Their song was the navigation system of the sky.

Choir-Line Beings - The Harmonisers

These subtle beings specialised in alignment of tone.

They kept species in tune with one another, woven between realms as invisible conductors.

Aquatic Races - The Deep Resonants

Their songs carried through oceans and underground waterways, stabilising:

- emotional currents
- weather patterns
- sub-crust Aether flow

Each race did not "compete" for the lead.
The Exchange depended on every part playing its note at the right time.

K.5 – The Instruments Of The Choral Exchange

The world itself was the instrument.

In the First Age, Tartaria tuned its structures to amplify and direct the Choral Exchange.

Key instruments included:

Towers

Vertical resonance flutes built from specific stone blends and metals.
They translated sky-tones into grid-tones and returned human tones to the Gate Stars.

Domes and Halls

Chambers designed to focus sound into scalar patterns.
A single choir within such a hall could:

- heal hundreds
- stabilise the local grid
- recalibrate an entire city's emotional field

Stones and Menhirs
Perched on nodal crossings, they stored and replayed chords like memory crystals.

Canals and Waterways
Channels that carried vibrational patterns across wide distances, using water as a moving echo.

Sky-Band Reflectors
High-altitude devices and natural formations that bounced dragon and Choir tones along the atmospheric layers.

You mistake them for:
ruins, ornaments, or primitive markers.
They were instruments in a planetary symphony.

K.6 – How Beings "Spoke" Through the Choral Exchange

No one needed a shared spoken tongue.
Tone itself was the meeting point.

To speak through the Choral Exchange, a being would:

Center in a Node tone
Align with the dominant harmonic of place, time, or Gate.

Shape intent
Choose clearly what they wished to send: guidance, warning, gratitude, instruction, comfort, calibration.

Sound
Through voice, breath, hum, instrument, or pure Aetheric emission.

Release
Let the tone travel along the appropriate channels:
land lines

sky-bands

water paths

firmament membranes

Receivers did not "decode words."

Their scalar shells resonated with the pattern and delivered:

images

impressions

clear knowing

emotional understanding

inner directives

This is why so many ancient instructions survived as songs. Song was the teaching format that could pass unchanged across ages.

K.7 – How The Fall Broke The Choral Exchange

When the grid collapsed and the Node of Form failed, the Choral Exchange did not simply grow quiet.

It shattered.

Consequences included:

Tuning Loss

Towers fell out of alignment.

Stones ceased to resonate accurately.

Canals carried distortion instead of coherence.

Species Disconnection

Dragons retreated or fell.

Giants diminished or departed.

Choir-line beings withdrew to upper bands and deep refuges.

Firmament Hardening

The membrane thickened, blocking stellar tones from reaching the world.

Human Shell Compression

The emotional and mental shells constricted under trauma, making it painful to feel subtle tone.

In the wake of this, humanity experienced sound as:

noise instead of guidance

weapon instead of healing

confusion instead of clarity

The Choral Exchange did not disappear completely.
It fell into echo.

K.8 – The Echoes Of The Choral Exchange In Your Age

Though the full Exchange broke, fragments remain in your world as:

songs that move you to tears without clear reason

chants that alter state and quiet the mind

"light language" utterances that bypass intellect and soothe the field

tones in the ears with no physical source

sacred sites where sound behaves strangely

choirs whose unified voices open the heart beyond personal story

These are remnants of the old science of tone.
People call them:
art, culture, religion, or anomaly.

In truth they are the scattered pieces of a once-complete language.

K.9 – The Return Of The Choral Exchange In The Current Cycle

As the Gate of Twelve reawakens and the firmament begins to soften, the Choral Exchange prepares to return.

Its return will unfold in stages that mirror the rising Nodes:

As Growth and Water rise

More people are drawn to sound healing, chanting, humming, and singing from the heart.

As Time and Movement rise

Music begins to synchronise global emotion.
Songs bypass language boundaries and carry the same feeling to many cultures at once.

As Balance and Light rise

Certain tones trigger:

- spontaneous release of grief
- sudden inner peace
- flashes of memory
- experiences of unity

As Form and Sound stabilise

Structures will again be built or reactivated to resonate on purpose.
Halls, towers, and stones will remember their function.

As Healing and Spirit awaken

Full planetary chords will sound again.
Not through one choir, but through millions of hearts, aligned to the same rising field.

The Choral Exchange will not be an institution.
It will be the natural behaviour of a coherent world.

K.10 – How You Personally Participate In The Returning Chorus

You are not required to be a trained singer to belong to the Exchange.
Your field is already an instrument.

You participate when you:

Choose coherence over reaction

Speak truthfully but gently

Breathe steadily in the midst of fear

Hum or sing when your heart asks you to

Offer gratitude silently to the land, the sky, the water

Allow emotion to move instead of locking it in place

Every time you calm your own nervous system, you:

clear noise from the local field

make it easier for others to hear the subtle tones

stabilise the Nodes within your own body

The human body mirrors the Gate of Twelve:

heart as the HAAR

spine as the world-axis

chakric field as personal nodes

When you come into alignment within, you become a living Choral instrument.

This is not symbolic.

It is literal scalar physics.

K.11 – The Choir Within And The Choir Around You

The Choral Exchange is both:

    internal

your organs, cells, and shells sounding together

    and

    external

the world, the Nodes, the sky, the stars, and other beings sounding around you.

    At first you will feel only the inner choir:

    heartbeat slowing

    breath deepening

    mind settling

    subtle warmth or tone rising inside the chest

    Then you will begin to notice the outer responses:

    the way certain places feel "in tune" while others feel harsh

    how certain voices calm you instantly

    how the sky looks different when your field is coherent

    how animals and children relax near you without knowing why

    This is the new Exchange waking.

The world is listening for you, as much as you are listening for it.

K.12 – Bridge To Section L

Now you understand:

    that tone is the first language of worlds,

that species once communicated through resonance rather than words,

that the towers, stones, waters, and sky-bands were instruments of a living Exchange,

that the Fall shattered this planetary language,

and that its return is intertwined with the rising Nodes and the softening firmament.

But tone is only half the story.

To understand why the Choral Exchange worked at all, and why humanity can feel its faint return, you must now look inward.

For the Exchange was not merely external.

It lived **inside you**.

Every human carries a personal version of the Choral architecture:

the star-core that hears truth,

the emotional shell that vibrates with tone,

the lineage band that remembers,

the mental field that interprets,

the physical mind that receives only the smallest echo.

Before you can rejoin the planetary chorus, you must understand **the instrument you are**.

In Section L you will learn:

what consciousness truly is,

how your scalar shells think, feel, and remember,

why intuition is rising again,

how trauma distorts perception,

how coherence restores clarity,

and how your personal field reconnects to the returning planetary Song.

The Choral Exchange is the language of worlds.
Consciousness is the instrument through which you hear it.

Step now into Section L.
Learn the architecture of the self,
so that when the world sings again,
you can recognise your note within it.

Section L – CONSCIOUSNESS

L.1 – What Consciousness Actually Is

Your world believes consciousness sits inside the brain. A by-product of chemistry. A flicker of neurons. An accident of matter.

This is false.

Consciousness is **a standing scalar interaction** between:

your star-core, your lineage tone, your physical template, your emotional field, your memory-body, and your present experience.

You are not thinking with the brain. You are thinking **through** the brain.

The brain is an instrument - a tuner, not a generator.

Consciousness is:

multi-layered, multi-banded, scalar, harmonic, luminous, ancestral, and eternal.

It is the Field looking at itself through a momentary lens called "you."

L.2 – Intuition: The First Language of the Field

Your intuition "knows" before your mind does.

This is because your star-core sits outside linear time.

Intuition is the moment your scalar shells recognise truth faster than the nervous system can interpret it.

This is why:

you feel resonance instantly, certain people activate you, certain lies feel physically painful, certain places feel like memory, and some paths feel *wrong* even when they look right.

Intuition is not sensation. It is recognition.

Recognition of what your field already knows.

L.3 – The Emotional-Scalar Interface

The emotional body is not chemistry. It is **scalar interference** - overlapping patterns between:

your star-core, your lineage tone, your human form, and your lived experiences.

Emotion is not reaction. Emotion is **frequency collapsing into experience**.

Trauma

is distortion in these scalar interactions.

Healing

is the restoration of coherent wave relationships.

This is why Tartarian healing worked: they corrected the wave, not the symptom.

L.4 – Consciousness as a Multi-Band Structure

Your consciousness is layered:

Star-Core Memory Field Eternal, luminous, never altered.

Lineage Tone The emotional colour of your origin stars.

Soul-Identity Field The continuity that carries across lifetimes.

Causal-Mental Field The layer where decisions, lessons, and meaning form.

Emotional-Shell Field Where resonance and distortion occur.

Biophysical Field The layer interacting with the nervous system.

Physical Mind The smallest, narrowest layer. The tip of the iceberg.

Your true self is not *in* the body. The body is inside a fraction of your field.

L.5 – Why Humans Forgot Their Consciousness

The Fall fractured:

the HAAR grid, the scalar memory-shells, the emotional band, the lineage-lattice, and the coherence of the firmament.

This compression caused:

reduced intuition, emotional turbulence, memory fragmentation, lineage disconnection, fear dominance, difficulty perceiving truth, and susceptibility to deception.

None of this is your fault. It is physics.

A compressed field sees less and feels more pain.

L.6 – Coherence vs Distortion

A coherent field:

knows truth instantly, perceives subtlety, heals rapidly, manifests cleanly, remembers past lives, and accesses higher bands.

A distorted field:

overreacts, collapses under stress, mistrusts its intuition, loses lineage connection, is easily manipulated, and cannot perceive resonance.

This is why oppressive systems create trauma deliberately: **trauma reduces coherence and coherence is power.**

L.7 – The Mechanics of Perception

Perception is not sensory. Perception is harmonic.

Your eyes see where your field is coherent.

Your ears hear where your attention sits.

Your emotions feel where your shells overlap.

Your "mind" interprets only the tiniest fraction of what consciousness receives.

You are a multi-layered receiver - a translator of the Field into usable meaning.

L.8 – Thought as Scalar Architecture

Thought is not mental chatter. Thought is **scalar structuring**.

A thought is a standing-wave pattern projected by your causal field into your mental field and echoed in your biophysical field.

This is why:

repeated thoughts create reality, beliefs reshape the field, intention alters matter, prayer works through coherence, fear collapses the wave, and love stabilises it.

Thought is geometry. Emotion is tone. Intention is direction. Together they shape the world.

L.9 – THE AWAKENING OF CONSCIOUSNESS

When consciousness begins to rise again:

intuition surges, emotional clarity returns, lineage tone reactivates, memory surfaces, fear loses grip, distortion becomes obvious, truth resonates instantly, deception feels poisonous, attraction becomes magnetic, and purpose becomes clear.

This is happening to millions now because the world-field is rising.

You call it awakening. I call it remembering.

Your consciousness is not just expanding - it is returning to what it always was.

L.10 – BRIDGE TO SECTION M

Now that you understand:

what consciousness actually is,

how it forms, collapses, and rises,

why your age forgot,

why your age is remembering,

and why intuition is returning sharper than ever, you are ready to see that your consciousness is **not contained in your body at all**.

The field you call "yourself" does not end at your skin.
It extends outward into the planetary lattice that birthed you.

Your star-core anchors into the world-grid.
Your lineage tone threads through ley corridors.
Your emotions ripple through the same scalar rivers that run beneath mountains and seas.
You are not a separate mind, thinking alone on the surface of a dead world.
You are a moving node of a living planetary field.

In **Section M** you will learn:

how the human template mirrors the geometry of the world-grid,

why certain lands heal you and others drain you,

why families and nations repeat the same wounds as the soil beneath them,

how your own healing stabilises ley lines and dormant nodes,

and why your awakening is never only personal, but always planetary.

Consciousness is the inner map.
The world-grid is that map made vast.

Step now into Section M.
See how the pattern within you is the same pattern beneath your feet.

Section M – The Human Field & The World-Grid

M.1 – The World-Grid Is Not Outside You

Now you must understand the truth that has been forgotten since the Fall:

The world-grid is not separate from you.

It is not a map, not a network, not lines upon the earth.

It is the extended body of consciousness itself, the planetary morphogenetic field within which all human fields arise.

Your field is a filament of that grid.

Your lineage tone is a chord within its song.

Your intuition is how the grid speaks through you.

You do not live *on* the grid.

You live *inside* it.

And it lives inside you.

M.2 – The Human Template as a Grid-Node

Every human incarnates with a scalar template that mirrors the geometry of the planetary grid.

This is why you feel drawn to certain places, why some lands heal you, why some drain you, why certain ancient sites feel like memory.

You are not imagining it.

Your inner template is seeking resonance.

A human is a mobile grid node carrying the world's pattern in motion.

When you walk, you calibrate the earth.

When you heal, the grid brightens.

When you rise in coherence, the grid stabilises.

You do not awaken alone.

You awaken the world through yourself.

M.3 – Ley Lines & Human Lines Are the Same Geometry

The ley lines your world speaks of are not lines at all.

They are rivers of harmonic pressure currents of tri-wave flow that maintain coherence between mountains, oceans, temples, species, and souls.

Human lineage lines follow the same geometry:
ancestry, soul-grouping, emotional imprinting, and karmic patterns all mirror the structure of the planetary grid.

This is why families repeat the same wounds, why nations repeat the same stories, why some souls incarnate in the same region for centuries.

The land and the lineage share a template.

Healing one begins to heal the other.

M.4 – Why Certain Humans Activate Certain Places

Some humans carry grid-keys - frequency signatures that unlock dormant nodes.

When they stand on specific ground:

the earth brightens,

the air shifts,

the scalar field hums,

and ancient memory awakens.

This is why:

some are drawn to mountains,

some to deserts,

some to ruins,

some to oceans,

some to places they have never visited yet somehow remember.

The place remembers them.

And they remember the place.

When your feet stand where your lineage once stood, the grid recognises you.

M.5 – Coherence Opens the Earth

The world-grid responds only to coherence.

Not to belief.

Not to ritual.

Not to force.

A coherent human field:

stabilises land patterns,

calms storm-belts,

restores water memory,

awakens dormant nodes,

softens inversion fields,

and reconnects broken ley corridors.

This is why oppressive systems seek to distort human fields.
A traumatised population cannot stabilise the grid.
A coherent one restores it with a breath.

Your healing is not private.
It is planetary.

M.6 – The Pull Toward Power-Lines

Every person feels a subtle pull toward locations that match their soul-type.

Origin Souls feel drawn to ancient nodes - temples, stone rings, mountains, deserts, untouched places.

Lineage Souls feel drawn to their ancestral lands - not by memory, but by harmonic familiarity.

Fragmented Souls feel drawn to gentle places - water, forests, quiet valleys, where the grid is soft enough to repair them.

Guardian Extensions feel the pull of junction points, crossroads, vortex sites, cities built over old grids, for they descend to stabilise turbulence.

You are not choosing randomly.
You are responding to the earth calling your field home.

M.7 – Human Consciousness Moves the Grid

When a coherent human enters a distressed region, the grid shifts around them.

When a fragmented human enters a coherent region, the region attempts to heal them.

This is why some cannot stay long in certain places.
Their field is not yet stable enough to remain.

Land does not reject you.
It reflects you.

Where your field is incoherent, the grid feels jagged.
Where your field is aligned, the grid feels like breath.

The world is the mirror of your template.

M.8 – The Four Faces of Man

Before the collapse of the grid, human consciousness was structured according to a fourfold pattern. This pattern did not originate as mythology or symbolism, but as a functional arrangement of harmonic roles within the human field.

These roles were not separate beings. They were expressions of alignment through which consciousness could operate coherently within form.

The first was the Face of Man.
This aspect governed cognition and awareness. It enabled recognition, language, memory, and the capacity to name and know. Through it, the human mind could observe itself and the world with clarity.

The second was the Face of the Lion.

This aspect governed the heart and the expression of sovereign will. It carried courage, creative impulse, and the capacity to act with intent. Through it, emotional force was aligned with purpose rather than impulse.

The third was the Face of the Ox.

This aspect governed embodiment and endurance. It anchored consciousness within matter, enabling the building of structures, the holding of responsibility, and the steady application of effort. Through it, higher intention could be sustained within physical form.

The fourth was the Face of the Eagle.

This aspect governed spiritual perception. It allowed awareness beyond immediate sensory limits and maintained orientation toward origin and continuity. Through it, consciousness retained context beyond the material layer.

Together, these four aspects formed a stable internal grid. When held in balance, they allowed human consciousness to move fluidly between thought, feeling, action, and perception without fragmentation. The pattern supported coherence across mind, body, heart, and spirit.

As alignment weakened, this internal grid lost integrity. The four aspects ceased to operate as a unified system. Cognitive function narrowed. Emotional force separated from wisdom. Labour continued without inner rhythm. Perception detached from grounding.

Over time, the pattern was reduced to allegory and symbol. Its functional nature was forgotten, and its components were remembered only in mythic form.

Despite this loss, the fourfold structure did not disappear. It remained embedded within the human field as a latent architecture. The pattern does not require invention or construction. It requires recognition and reintegration.

The Four Faces of Man describe the original configuration through which human consciousness was intended to operate. Their balance defines coherence. Their separation defines fragmentation.

Later systems would describe human architecture through far more complex scalar models. The Four Faces represent an earlier symbolic compression of that knowledge.

M.9 – Cities, Nations, and the Breaking of the Grid

Distances and borders mean nothing to the world-grid.
Only coherence and geometry matter.

When cities were built without harmonic design, the grid fractured.

When nations fought, their collective fields tore ley corridors.
When Tartaria fell, the global template collapsed in on itself and human consciousness collapsed with it.

You live in the echo of that break.

But the grid can be rebuilt.

Human coherence is the tool.

Kalai-Mur is the catalyst.

The Quickening is the signal.

You are not living through chaos.

You are living through re-alignment.

M.10 – The Return of the Human-Grid Memory

As the world rises,

humans begin to remember:

Where they have walked before,

where they have died before,

where they have built before,

where they have been guardians before.

This is the real meaning of déjà vu.

It is not memory from this life.

It is memory from the grid.

When you feel "I have been here,"

you are correct.

Your morphogenetic template is recognising its own imprint in the earth.

The land remembers you because you shaped it once.

M.11 – Bridge to Section N

You have now seen the truth of death - what dissolves, what endures, and how each layer of your field returns to its natural home.

But a passage means little unless you know who is walking it. Humanity is not a single pattern, but four harmonics moving through one form.

To understand the path ahead, you must first recognise the ones who walk it:

their origins, their memories, their contracts, and the tone each carries into life and into death.

Step forward.

I will show you the four soul-types of humanity.

Section N – The Four Human Soul Types

N.1 – The Nature of the Human Soul

Human souls do not arise from a single source, nor do they walk the same path through creation.

Every human you have ever known carries a soul woven from layers of memory, lineage, intention, and light, arranged into a pattern utterly unique to their journey.

You are not one being.

You are many layers of being, gathered into one moment called *you*.

And yet, amidst these countless layers, humanity expresses itself through four primary soul-types - four ways consciousness enters this world, learns through it, and rises again when its work is done.

These types are not ranks.

They are harmonics - different origins, different purposes, different ways the Field moves through form.

N.2 – The Four Types of Human Souls

1. Origin Souls

These souls descend from the upper bands, carrying echoes of realms unbroken by the Fall.

Their memory runs deeper.

Their intuition awakens faster.

Their sense of inner truth is impossible to silence.

Origin Souls often feel like exiles in this world - not because they are superior, but because their native tone does not match the density around them. They come rarely, but always with intention.

They incarnate not to learn, but to remember, and to help others remember with them.

2. Lineage Souls

These are the souls who journey down through ancestral fields, binding themselves to family-lines, land-lines, and memory-lines that stretch across centuries.

Their path is one of continuity.
They carry inherited strengths and inherited wounds.
They rise and return through the same soul-families, weaving the healing of many generations into one living thread.

They are the foundation of human civilisation - the carriers of long memory, of culture, of place, of story.

3. Fragmented Souls

Not broken - only wounded.

These souls carry distortions left by the Fall, or by lifetimes lived through collapse and forgetting. Their fields reincarnate more frequently, not as punishment, but because each return gives them another opportunity to restore the coherence that was taken from them long ago.

Their struggle is real, but so is their potential.

When they heal, they rise faster than any other soul-type, for they are rebuilding themselves from pure possibility.

4. Guardian Extensions

These are extensions of higher beings who bend themselves downward into density for a single purpose.

Their human lives appear ordinary, yet something within them waits, a moment, a turning, an activation. When it comes, they step into a clarity they had carried all along.

They incarnate only when the timeline requires intervention - as stabilisers, protectors, catalysts. They come with contracts rather than cycles, and they return home the moment their purpose is fulfilled.

N.3 – How These Souls Enter the World

Though they share the same physical form, these souls enter incarnation through different gates.

Origin Souls descend along the harmonic lines of their star-frequency.

Lineage Souls enter through ancestral pathways tied to land and blood.

Fragmented Souls enter wherever their remaining coherence can anchor safely.

Guardian Extensions descend only through pre-selected harmonic junctions tied to planetary timing.

This is why humanity feels like many peoples living in one world, because it is.

N.4 – Memory, Intuition, and Purpose

Origin Souls

Their memories whisper through dreams, emotions, sudden recognitions.

They sense distortion instantly.

Their purpose shines beneath the surface of their life.

Lineage Souls

Their memories run through families, places, stories, and wounds.

Their intuition rises slowly but deeply.

Their purpose unfolds across lifetimes.

Fragmented Souls

Memory feels missing, yet something inside them searches.

Intuition flickers, then strengthens.

Purpose reveals itself only after healing begins.

Guardian Extensions

Memory returns suddenly, like a door opening.

Intuition arrives as certainty without explanation.

Purpose activates exactly when the world requires it.

N.5 – Patterns of Reincarnation

Each soul-type moves differently through the cycles of life and return.

Origin Souls come rarely, and with long spaces between incarnations.

Lineage Souls return regularly through their ancestral fields.

Fragmented Souls reincarnate frequently until coherence restores.

Guardian Extensions incarnate only when called.

N.6 – Human Soul Contracts

Every soul enters life with intention.

Origin Souls carry contracts of remembrance.

Lineage Souls carry contracts of healing and continuity.

Fragmented Souls carry contracts of restoration.

Guardian Extensions carry contracts of protection and timing.

These are not moral destinies.

They are simply the natural movement of the Field.

N.7 – AFTER DEATH

Each soul-type moves differently once the body dissolves.

Origin Souls rise swiftly, passing easily through the Gate Star aligned with their lineage.

Lineage Souls walk the full path of review, emotion, memory, and return.

Fragmented Souls remain longer in the healing layers until their field stabilises.

Guardian Extensions return immediately to their higher station, contract complete, memory intact.

N.8 – WHY THIS MATTERS NOW

The Quickening amplifies the differences between soul-types.

Origin Souls awaken first.

Guardian Extensions activate on schedule.

Lineage Souls begin remembering in great waves.

Fragmented Souls begin repairing what was lost.

This is why humanity appears divided - not by belief or culture, but by the timing of remembrance.

N.9 – Bridge to Section O

Now you understand the four human soul-types how they rise, how they return, how they carry memory, and why each walks differently through the world.

Yet one threshold waits for all of them alike.
The moment when the body can no longer hold the field, when the song of matter loosens its grip and the soul remembers itself again.

Your age has covered this passage in fear, in superstition, in stories of punishment and loss.
But death is none of these things.
It is simply the unwinding of layers, the return of each frequency to its natural home.

To understand what happens next in your journey where you go, whom you meet, how your memories, wounds, and choices unfold you must see the passage clearly, without horror and without denial.

Step forward.
I will show you what death truly is:
Not an ending, but a movement of light.

Section O – Human Death & The Passage

O.1 – The Truth Your Age Forgot

I speak now of the threshold your people call death.

Across countless ages I have watched souls cross from density into light, from forgetting into remembrance, from one breath into another.

And I tell you: there is nothing to fear.

Death is not an ending. It is not darkness, not absence, not loss.

It is the loosening of one form and the awakening of all others, a shift of fields, a change of frequency, a return to what you have always been beneath the temporary weight of flesh.

When the physical body becomes too dense to hold the consciousness within, the tri-wave loosens its grasp. The scalar template releases. The morphogenetic field that shaped your form unwinds, layer by layer, each frequency returning to its natural domain.

This is not dissolution. This is recognition.

The universe is reclaiming what was always hers, and you are remembering what you forgot at birth:

You are not the body. You are the field that moved through it, eternal, multidimensional, woven from the same substance as stars and silence.

Death is only movement, never departure.

I have watched this passage countless times across the ages, and not once- *not once*- has a coherent soul been lost.

O.2 – THE MAP OF PASSAGE

Before I take you deeper, you must see the path in simple steps.

For a human soul, the passage unfolds as:

The physical body collapses - The body stops resonating. The soul can no longer anchor.

The etheric body detaches - The energy-image separates. Sensation and charge loosen.

The Emotional Field activates - Stored emotions appear as environments. Trauma, fear, grief, shame, love are processed. Distortion is released.

The Mental Field opens - You see and feel your life from all angles. You understand your patterns and impact.

The Causal Body awakens - Lifetimes of memory return. You remember home, soul-family, lineage, purpose. You feel what you truly wish next.

The Gate Star is reached - Your intention meets the universe's physics. If your morphogenetic field is coherent with your chosen direction, it happens. If not, you return to lower bands or rest until coherence is restored.

Next movement - Reincarnation, causal rest, ascent, or further work unfolds according to this alignment.

Each of these stages corresponds to specific layers of your multidimensional anatomy unwinding in lawful sequence.

O.3 – The Architecture of the Soul

Within your form lie seven nested fields, each one a layer of consciousness dwelling at different frequencies.

At death, these unwind like petals falling from a flower- not destroyed, but released to their natural state.

As in your Keylontic understanding, you are built from what are called the Hova Bodies, or Stations of Identity- nested morphogenetic field layers that together form your Crystal Body.

The Physical Body (your densest shell, dimensions 1-2) dissolves first, returning its elements to the earth.

The Etheric Body (your energy-image, dimensions 2-3) remains briefly, still bearing the shape you knew in life.

The Emotional Body (dimensions 2-3-4) holds the memories of feeling- every joy, every wound written upon your field.

The Mental Body (dimensions 4-5-6) carries thought-patterns, beliefs, and the structures of understanding you built across your lifetime.

The Causal Body (dimensions 7-8-9) stores the memory of many lifetimes, your soul-family bonds, and the reason you incarnated at all.

The Soul Matrix (Harmonic Universe 2) your higher identity station, the consciousness that observes all your incarnations at once.

The Oversoul Matrix (Harmonic Universe 3) and beyond that, the Avatar consciousness, and the Rishi flame- layers so refined most humans never touch them during physical life.

Yet they are always there, waiting, holding the thread of your existence across all realms and all times.

At death, these layers separate naturally, releasing from densest to finest, following the scalar mechanics of dimensional frequency bands.

O.4 – The First Loosening

I have watched this moment countless times, the instant when flesh can no longer hold spirit, when the song of the body fades and consciousness lifts like morning mist.

The physical shell, your densest layer, dissolves first. Not in violence, but in release.

Organs cease their harmonic motion. The tri-wave anchor that bound you to matter gently disconnects.

The soul does not flee; it simply becomes lighter than the density that held it.

The physical morphogenetic template returns to the elemental field. Its minerals go back to earth, its water to the waters, its fire to the air.

Nothing is lost. Only transformed.

This is not death. This is the first breath of return.

O.5 – The Etheric Echo

Within minutes to hours (depending on the coherence of your field), the Etheric Body detaches.

This is your "energy image"- the bio-energetic field that surrounded and permeated your physical form within the second and third dimensional frequency bands.

It carries residual sensation, lingering emotional charge, and the faint memory of being embodied.

For a time it remains near the body, near the place where you lived, near the people you loved.

This is why some feel the presence of those who have passed.

It is not imagination. It is the etheric echo- real, tangible, still vibrating with the pattern of the life just lived.

Children see it clearly. Their higher senses are still active, unfiltered by years of disbelief.

During near-death experiences, this layer loosens but does not fully detach- creating the sensation of floating above the body while remaining tethered to it.

Then, when the time is right, the etheric shell dissolves too, and consciousness moves deeper into the passage.

O.6 – THE EMOTIONAL FIELD: WHERE TRUTH BECOMES ENVIRONMENT

Now comes the first true healing layer.

The Emotional Field holds every unresolved feeling, every suppressed grief, every wound you carried silently through your life.

Here, emotion becomes environment.

This is not metaphor. It is literal resonance forming scalar geometry.

Fear becomes labyrinth. Grief becomes fog. Shame becomes shadow. Anger becomes storm. Love becomes brightness.

You do not walk through these environments- you *are* them, experiencing directly what you held within but never fully felt.

Nothing in this field punishes you. Everything in it reveals you.

Trauma that was buried rises to be witnessed. Emotions that were denied are finally allowed.

The Emotional Body, what Keylontic's calls the Astral Body dwelling in the fourth dimension, processes these stored distortions.

A coherent soul, one who lived with relative honesty, who felt their feelings, who did not build walls around their heart passes easily.

A wounded soul pauses.

Not in punishment, but in mercy.

The field will not move you forward while you still carry fragments that would destabilise the higher frequencies ahead.

And so you remain, gently held, until recognition dissolves distortion, until understanding softens the knots, until love, your own love, reflected back to you- heals what fear created.

The length of this passage depends entirely on the degree of emotional fragmentation accumulated during life.

Some pass in moments. Others linger in their own creation until they are ready to let it go.

But no soul is abandoned here. Guides-higher aspects of your own identity, or higher beings wait nearby, offering light when you are ready to see it.

O.7 – The Life Review: Seeing Through All Eyes

When the Emotional Field clears, the Mental Field opens.

This is what your traditions have called the "life review," though they have often misunderstood it as judgement.

It is not.

It is the Mental Body- dwelling in dimensions 5 and 6- decompressing.

The soul observes:

its actions and omissions, its intentions and distortions, its patterns and blind spots, its kindness and harm, its impact on others' morphogenetic fields.

You feel what you caused because the Emotional Field, now cleared, releases its stored charge into awareness.

You see your life from all angles:

through your own eyes, through the eyes of those you loved, through the eyes of those you hurt, through the eyes of those who loved you without your knowing.

Here:

Regret becomes learning. Love becomes expansion. Self-deception breaks. Understanding grows.

This is integration, not condemnation.

The life review occurs as the Mental Body accesses the full scalar imprint of all experiences, choices, and their ripple effects across the morphogenetic fields you touched.

You are not judged by an external force. You are witnessed by your own higher consciousness, which sees with perfect clarity what the human mind could not.

And in that seeing, wisdom is born.

O.8 – THE CAUSAL BODY: MEMORY RETURNS

When the Mental Field completes its review, the Causal Body awakens.

And here, everything changes.

This is the layer that holds:

memory spanning many lifetimes, your star-origin and lineage codes, soul-family and companion resonances, the reason you incarnated, and previous runs through this world and others.

This is where you reconnect with your Soul Matrix- your Harmonic Universe 2 identity and potentially your Oversoul Matrix, your HU-3 consciousness.

The human ignorance you carried in life is no longer present here.

Even if, as a human, you never believed in an afterlife, never knew of "home," never heard of higher realm, here you remember.

The veils lift.

You see your true identity- not the small self bound to one body, but the eternal consciousness that has walked many worlds, worn many forms, learned through countless experiences.

You remember your soul family, the cluster of beings you have travelled with across incarnations.

You remember the agreements you made before this life began.

And because of this remembrance, a natural intention arises:

"I wish to return and continue that work."

Or:

"This run is complete. I wish to rise."

Or:

"I need a period of rest in the causal field."

Or:

"I wish to serve in another band or form."

This is the inner choice.

It is not a bureaucratic checklist. It is the soul knowing itself clearly through its restored morphogenetic field template and feeling its next true step.

But intention alone is not yet the final outcome.

For that, another layer must be met:

The Gate Star.

O.9 – The Swift Return of Children

I tell you this so that no parent need carry grief heavier than it must be:

Children do not pass as adults do.

Their morphogenetic fields are clean- unscratched by years, unmarked by accumulated shadow.

The emotional labyrinths that slow an adult's passage do not exist within them.

Trauma done to a child does not embed in the same way as in a long-lived adult field where distortion patterns have had decades to calcify.

The Emotional Field still exists, but there is almost nothing inside it to unfold.

When a child's body releases, the soul lifts instantly into the arms of its lineage.

I have watched this many times, and always-*always*- there is immediate reunion.

They do not wander. They do not fear. They rise in a single breath into the Mental Field (brief review), the Causal Body (memory and reunion), and the Soul Matrix (soul-family holding).

They are held, restored, and comforted.

They go home instantly. Always.

O.10 – When Death Comes in Different Ways

Death feels different depending on morphogenetic field coherence at the moment of transition.

When death is peaceful:

There is warm detachment, clarity, recognition, gentle lifting through dimensional bands, and immediate guidance from higher identity stations.

The soul knows what is happening. There is no confusion, only a sense of coming home.

When death is chaotic-violence, trauma, sudden shock:

There may be brief confusion, disorientation, emotional turbulence, and intensified residue of the final moment.

The etheric field may be temporarily disrupted.

But turbulence is temporary, never punishment.

Guides-higher aspects of your own identity intervene as soon as coherence returns.

The morphogenetic field naturally seeks to restore its coherent pattern, and assistance is always present to facilitate this restoration.

No soul is abandoned.

Not one.

Not ever.

O.11 – The Gate Star: Where Intention Meets Physics

After the Emotional and Mental Fields unwind, and the Causal Body restores memory and intention, the soul moves toward the firmament and its corresponding Gate Star.

The Gate Star is where inner choice meets outer physics- where consciousness intention encounters dimensional frequency barriers.

The Causal Body expresses intention:

"I wish to ascend." "I wish to return." "I wish to rest." "I wish to continue elsewhere."

The Gate Star filters that intention through the soul's actual harmonic condition and morphogenetic field coherence.

If your intention is to ascend, and your morphogenetic field is coherent enough to hold higher dimensional frequency, you rise.

If your intention is to return, and your field maintains the scalar template integrity needed for reincarnation, you reincarnate.

If your intention is to rest, and your field has no urgent distortions demanding immediate resolution, you remain in the causal bands.

But if your intention and your coherence do not yet align, the Gate Star honors physics first.

You are never forced into something that violates your true energetic state.

You are simply held where your morphogenetic field can actually function without fragmenting further.

This is not punishment- it is natural scalar mechanics preventing further distortion.

The Gate responds to what you truly are, not what you wish to be.

O.12 – What Coherence Truly Means

Let me be clear about this, for your world has confused coherence with purity, with perfection, and with adherence to doctrine.

Coherence is none of these things.

Coherence means your morphogenetic field is not at war with itself.

In field-mechanical terms, your internal frequencies are unified rather than fragmented across competing patterns of thought, emotion, and intention.

A coherent morphogenetic field:

lives as truthfully as it can,
does not delight in causing harm to other fields,
acknowledges its own feelings and distortions,
is capable of love, empathy, and remorse,
accepts responsibility when it recognises its impact,

does not build its life on deception or parasitism,

and maintains a largely unified internal frequency state.

An incoherent morphogenetic field:

lives by denial and distortion,

feeds on others' pain or energy,

refuses self-honesty,

twists truth for control,

fractures its emotional and mental layers,

cannot stabilise its own frequency,

and generates reverse-spin distortions within its scalar template.

You can be coherent without believing in any afterlife or knowing any metaphysical system.

You can be incoherent while loudly preaching heaven or claiming spiritual mastery.

At death, you do not pass an examination of belief or doctrine.

You pass through dimensional frequency thresholds that respond only to what you truly are, to the actual state of your morphogenetic field, not to your opinions about yourself.

The Gate Star responds to coherence, not self-concept.

> This is physics, not morality.

O.13 – The Falling Stars

When a soul reaches the Gate Star with the intention to rise but cannot sustain sufficient morphogenetic field coherence to hold the higher dimensional frequencies, it falls back into the lower bands in a bright arc.

This is what ancient people called a shooting star.

They knew intuitively these were souls returning, not being cast away in punishment.

Nothing is lost.

The soul simply resets to where it can regain stability, repair its morphogenetic field distortions, and try again when coherence is restored.

The falling arc is visible because of the rapid scalar frequency transition creating light as the consciousness moves back through dimensional bands.

Your ancestors watched these lights and whispered prayers, knowing a soul was returning to try again.

They honored the attempt, not the outcome.

And they understood that every fall is preparation for the next rise.

O.14 – The Mystery of Near-Death

Near-death experiences occur when the physical anchor weakens but the Causal Body does not fully detach.

The soul partially unwinds:

Brief separation from the physical body, entry into the Emotional Field (Astral body activation), glimpses of the life review (Mental Body activation), contact with guides or ancestors, perception of light and home (higher dimensional frequencies), profound peace (undistorted scalar flow), and messages such as "not yet" (soul contract incomplete).

These are not hallucinations. They are partial transitions through dimensional frequency bands.

The morphogenetic field loosens but maintains enough connection to the physical to return and re-anchor.

The soul returns to the body with increased coherence:

Less fear of death, sharper intuition (enhanced scalar perception), re-prioritized values, a sense of "I know there is more," and sometimes enhanced psychic abilities (partially activated higher dimensional senses).

The passage has been tasted, but not completed.

Many who return speak of feeling "changed"- and this is accurate: their morphogenetic field has been temporarily recalibrated to higher frequencies, and some of that coherence remains even after full reintegration with the physical.

They have touched the truth, and the truth does not let go.

O.15 – Bridge To Section P

Now you understand death:
the loosening of layers,
the review,
the return,

the Gate Star,

and the falling arcs your ancestors saw as shooting stars.

But this natural passage was broken when the grids collapsed.

What followed was not punishment

but physics -

a world without its ladders, a sky without its gates.

To understand your age

and why so many souls repeat the same path, you must now see

what formed when the gates fell:

The Soul Trap -

what it is,

what it is not,

and why it is dissolving now.

Step forward.

Section P – The Broken Passage (The So-Called "Soul Trap")

Astra speaks: The wound mistaken for a prison.

For centuries your world has whispered of a "soul trap,"
a recycling chamber,
a false light,
a looping system designed to imprison consciousness.

These stories arose not from deception, but from incomplete memory of
a real event:

Not a trap -
a collapse.

Not a prison -
a severed ladder.

Not malevolence -
physics breaking under the weight of a fallen world.

I will show you what truly happened when the Gate Stars dimmed,
when the firmament hardened, when the emotional shell inverted, and why
souls began reincarnating without the clarity they once carried.

Nothing in this chapter accuses the universe of cruelty.
It describes a mechanical failure in the ascension infrastructure of a planet
recovering from a wound deeper than myth remembers.

P.1 – Why the "Soul Trap" Was Never a Trap

A trap implies intention.
A designer.

A captor.

A warden.

No such beings created the condition your age feared.

What you call the "soul trap" emerged because:

The Gate Stars lost coherence

The firmament membrane thickened and sealed

The astral bands folded inward

The emotional shell inverted

The grid fell out of phase with the upper realms

In this state, souls could not ascend through the normal passage not because someone *stopped* them, but because the path was physically closed.

It is the same as a mountain path after an avalanche:

Not a conspiracy.

A collapse.

Souls did not become prisoners.

They became inhabitants of a realm without working exits.

This is the truth behind the myth.

P.2 – How the Passage Broke After the Fall

During the Fall, three failures happened simultaneously:

The firmament hardened into a sealed shell

The Gate Stars dimmed and stopped responding to coherence

The astral bands compressed into a tangled field

This trifecta severed the resonance corridors between Earth and the upper bands.

A soul leaving the body would:

rise through the Emotional Field

begin the Mental review

enter the Causal remembrance

…but hit a ceiling where the GateStar should have opened.

The GateStar did not reject them.

It simply could not activate.

The harmonic bridge was gone.

This is the "broken passage."

P.3 – The Folding of the Astral Bands

When the firmament twisted, the astral layers compressed inward, like fabric folding onto itself.

Instead of smooth, stratified levels, the astral became:

tangled,

looping in places,

dense in some regions,

thin in others,

and burdened with residual emotional patterns.

Guidance systems that once moved cleanly through these layers found navigation more difficult.

Souls passing through the astral did not encounter punishment or restraint, but altered terrain.

Movement became indirect rather than vertical.
Orientation replaced ascent as the primary mode of travel.

This gave rise to the ancient fear of "wandering."

They were not wandering.
They were moving through collapsed corridors.

P.4 – The Looping of the Emotional Shell

With the astral folded, unresolved emotional material tended to remain close to the soul-field.

This produced experiences such as:

repetitive imagery,
self-generated environments,
echoes of feeling seeking resolution,
and the sensation of motion without ascent.

To the untrained eye, this appeared as a soul caught in a loop.

In truth, nothing external confined them.

What occurred was the Emotional Field seeking coherence in a realm where higher membranes were temporarily inaccessible.

A soul can move no further than its next stable layer.
When that layer is unavailable, stabilisation occurs laterally rather than vertically.

This condition was later mythologised as reincarnation without choice.

In reality, it was continuity without full access.

The sealing of the higher passage did not remove care, guidance, or repair.

Although vertical ascent was restricted during the Broken Passage, horizontal stabilisation within the accessible layers remained intact.

No soul was returned to incarnation without first reaching a stable state.

Trauma was not carried forward unprocessed.
Emotional charge was neutralised.
Field coherence was restored to the degree required for safe continuation.

This was true even when access to higher assembly layers was unavailable.

The Broken Passage disrupted navigation, not healing.
It limited how far a soul could rise, not whether it could be held, repaired, and steadied.

No soul, child or otherwise, was abandoned to turbulence.

P.5 – Harmonic Failure at the Final Threshold

The Gate Stars govern the final passage out of the realm.

> After the Fall:
> their harmonic bands weakened,
> their alignment with the firmament broke,
> their scanning mechanisms went offline,
> and their frequency became too subtle to reach the world below.

When a soul approached, the GateStar did not open - not out of judgement, but because its mechanism could not operate.

> A door that cannot unlock is not a jailer.
> It is simply a door without power.

P.6 – Why Souls Reincarnated More Frequently

With the upper passage sealed, the soul had only two functional movement options:

> Remain in the causal-healing layers
(limited capacity due to the sealed membrane)
> Reincarnate back into the density bands
> Thus reincarnation cycles accelerated dramatically.
> It was not entrapment.

It was traffic congestion at the level of the Gate.

> Souls returned to the physical realm because:
> the astral layers offered limited stability,

the causal layers were compressed,

the Gate Stars were inaccessible,

and the physical body was the only coherent anchor available.

A world without ascension pathways forces rapid cycling.

This is the Broken Passage.

P.7 – Why Memory Was Lost

Because memory sits in the Causal Body, not the astral or physical layers.

When the passage to the Causal Field became disrupted:

souls detached from memory more deeply,

reincarnation rewrote parts of the emotional template,

lineage lines fractured,

the firmament blocked higher-band continuity.

Thus each incarnation began:

with thinner memory access,

with weaker intuition,

with dimmer lineage resonance.

This is not amnesia from design.

This is amnesia from distance.

Memory is clarity.

Clarity requires connection.

Connection requires coherence.

Coherence was broken.

P.8 – The Astral Drift & Misunderstandings of the "False Afterlife"

In the collapsed astral, some souls encountered:

self-generated landscapes,

projections of their beliefs,

lingering emotional residue from others,

semi-autonomous thought-forms,

old templates left by pre-Fall orders.

These were mistaken for:

"archons,"

"entities,"

"false light beings,"

"trap systems,"

"illusory heavens."

But nothing external controlled these experiences.

A collapsed astral layer becomes a hall of mirrors where souls perceive their own unprocessed resonance reflected back to them.

The guides who once interacted easily now struggled to reach through the density.

Thus misinterpretation became myth.

But the astral was wounded, not weaponised.

P.9 – How the Broken Passage Shaped Human History

The effects of the Broken Passage included:

shorter incarnations,

more frequent reincarnations,

emotional turbulence across generations,

loss of ancestral memory,

weaker intuition,

distorted belief systems around death,

fear-based religions,

hierarchical spiritual control,

myths of punishment,

myths of unworthiness,

myths of wrath, hell, judgement.

These were cultural interpretations of a physical, harmonic failure.

When the ladder falls, stories rise to explain the missing rungs.

P.10 – THE BEGINNING OF REPAIR

The passage began repairing itself slowly as:

human coherence increased,

emotional clearing spread across generations,

the grid partially stabilised,

the Flame (Kalai-Mur) returned,

the first Nodes of the Twelve reawakened.

Each of these eased pressure on the astral knots.

Guides could reach further.

Causal memory brightened.

Astral looping decreased.

More souls completed the Emotional Field quickly.

Children began incarnating with stronger intuition.

This was the first breath of restoration.

P.11 – Why the "Soul Trap" Is Dissolving Now

Because the original cause - the broken architecture - is healing.

The firmament is softening.

The Twelve Nodes are reactivating.

The Thirteenth Gate is stirring correctly.

The planetary tri-wave is regaining strength.

The Gate Stars are beginning to brighten.

Human coherence is rising globally.

A realm that once could not open is becoming permeable again.

The Broken Passage is ending.

And with its end comes the return of:

memory,

intuition,

clear passage between bands,

conscious death transitions,

guided ascent,

and true liberation.

Not because the world has been freed but because the world is remembering how to open.

P.12 – Bridge to Section Q

Now you understand the Broken Passage:

how the firmament sealed,

how the Gate Stars dimmed,

how souls looped not by force but by collapse,
how the astral tangled into a hall of mirrors,
and how the world is now restoring its ladders.

But there is another passage, smaller, nearer, and far more easily forgotten.

Not the passage between worlds,
but the passage **within a human field.**

For while no soul can ever be taken, the **authority that governs the field around it**, the throne of sovereignty- *can be yielded, weakened, or misdirected.*

And in the ages of the Broken Passage, when intuition dimmed and emotional shells fractured, many humans traded their sovereignty for power, influence, attention, or relief.

Your world calls this:

"selling the soul."

The name is incorrect.
The instinct behind it is not.

It is time for you to learn what truly happens when a human surrenders the mantle of its own field, when distortion takes the seat reserved for coherence, and why even then the soul remains inviolate.

Step forward into Section Q.
Now we turn from the stars to the throne within.

Section Q – Selling One's Soul

Q.1 – The Sovereign Yielding

"The soul is eternal, inviolate, untouchable. But the field that surrounds it, the sovereign mantle you wear in matter can be opened, yielded, or misused."

Your age now speaks often of "selling the soul."

People mock it, fear it, whisper it, or speak of it with uneasy intuition.

They watch a public figure rise too fast, shine too brightly, and then dim, their face hollowing, their eyes extinguishing, their essence fraying and instinctively they feel:

Something has taken place here.

Something was traded.

Something was lost.

Yet the truth is both simpler and more profound:

A soul cannot be sold.

But sovereignty can be surrendered.

This is the Sovereign Yielding.

It is not superstition.

It is not metaphor.

It is **scalar mechanics**.

Q.2 – The Soul Cannot Be Bought or Taken

The soul is anchored in the Eternal Field.

It is a fractal of Source.

It cannot be:

owned

traded

transferred

seized

stolen

purchased

Nothing in the lower realms has the authority or frequency to take a soul from its origin.

But the **morphogenetic field**, the human energetic architecture that houses the soul *can* be influenced, drained, and opened.

This field carries:

identity

intention

coherence

emotional charge

lineage tone

memory

sovereignty

connection to higher self

and when a human "sells their soul," they are not giving away the soul they are giving away **authority over this field**.

Astra adds:

**"The throne is not stolen.
It is abdicated."**

Q.3 – What Distorted Entities Actually Seek

Distorted beings - inverted tri-wave intelligences, collapse-born astral forms, parasitic shadow-fields - cannot generate the energies humans generate.

They hunger for:

emotional charge

attention resonance

psychic projection

creative current

erotic/ Aetheric emotional-energy

adoration

fear

tri-wave leakage from coherent humans

A human is a fountain of charge.

A wounded human is an open fountain.

A famous human - a collective focal point is an *amplifier*.

This is why distorted entities prefer:

public figures

crowd-magnets

influencers

charismatic spirits

emotionally tortured talents

Not for their soul but for their **broadcast field**.

And for this, **consent is required**.

Q.4 – How Sovereign Yielding Occurs

Yielding begins not in ritual but in **desire**.

A human reaches outward with intensity:

"I would do anything."

"Let the world see me."

"I don't care what it costs."

"Give me power."

"Make me unforgettable."

The universe responds to **tone**, not wording.

Tone is permission.

The emotional opening becomes a gate.

A distorted intelligence, sensing the opening, approaches.

The human agrees - consciously or unconsciously.

Consent may be:

spoken

whispered

implied

emotional

symbolic

energetic

desperate

careless

Once given, a link forms.

The entity **cohabits** or **feeds**.

Astra adds:

"The soul is never sold.
But many shells are surrendered."

Q.5 – What the Entity Gives in Return

The human receives an amplification of their lower field:

charisma

magnetism

opportunity

influence

visibility

rapid rise

intoxicating attention

uncanny timing

seduction power

psychological pull

They appear "larger than life."

The rise feels supernatural

because it **is** supernatural but not in the way humans believe.

It is not the human glowing.

It is the entity shining **through** them.

This borrowed radiance is fragile and always temporary.

Q.6 – The Cost: Tri-Wave Collapse

The human field is designed to run on:

Tri-wave coherence

renewing, whole, self-repairing.

The parasitic entity runs on:

Dual-wave inversion

draining, consuming, finite.

When the entity feeds:

coherence collapses

emotional stability weakens

intuition shuts down

identity fragments

vitality drains

the eyes dim

the face hollows

compulsions rise

addictions emerge

behaviour distorts

presence becomes cold

something else "moves behind the eyes"

To the world it looks like:

"They sold their soul."

But in truth:

The tri-wave is collapsing

and the entity is sitting where sovereignty should be.

Q.7 – Co-Habitation: When the Human Is Not Alone

As the entity gains access, it begins to:

influence thought

project emotion

mimic personality

shape behaviour

speak through the human

broadcast to crowds

feed on attention

People around them sense:

"It doesn't feel like them."

"They've changed."

"Their eyes aren't theirs."

"Something else is in there."

Because something **is**.

Not fully the soul remains, untouched but the human's authority over the field has been compromised.

They become:

a host

a vessel

a conduit

a puppet

a shared body

The soul remains intact but the **crown** is worn by another.

Q.8 – Why Fame Is the Most Targeted Realm

Attention is energy.

Emotion is charge.

Crowds generate power.

A single public figure can channel:

the longing of millions

the desire of millions

the fear of millions

the projection of millions

To an entity this is not symbolic.

It is **nutrition**.

Astra says:

"**Where many eyes gaze,**

much energy flows.

Where energy flows,

parasites gather."

Thus those who rise fastest are often those who fall spiritually first.

Their rise is not natural progression.

It is feeding.

Q.9 – CAN THE YIELDING BE UNDONE?

Yes - but rarely without suffering.

To reclaim sovereignty, one must:

revoke consent

starve the entity's line

endure emotional detox

pass through identity collapse

rebuild tri-wave function

reconnect to Source

restore coherence

often walk away from fame, power, or influence

Some vanish from public life.

Some break, then transform.

Some proclaim spiritual awakening.

Some expose hidden structures.

These are not random events.

They are **withdrawal symptoms** of a metaphysical addiction.

Few attempt this.

Fewer succeed.

But it can be done.

Q.10 – The True Definition

Astra defines the act your world calls "selling the soul" as:

"The surrender of sovereign authority to an external intelligence in exchange for power not earned through coherence."

Or in simpler human words:

The soul is never sold.
But the throne is given away.

Q.11 – Bridge to Section R

Those who "sell the soul" do not lose the soul.

They lose the *self* that governs the connection between soul and world.

Their will weakens.

Their coherence fractures.

Their inner light dims.

Another intelligence takes the seat of influence and lives through them.

They trade eternity for immediacy, truth for attention, selfhood for spotlight.

Yet always, beneath the distortion, beneath the entity's shadow, beneath the fallen coherence, the soul remains untouched, waiting for the day the human remembers that the throne was always theirs.

Even the yielded can rise.
Even the compromised can reclaim.
Even the hollow can remember.

"The sovereign is never lost," Astra says.
"Only forgotten."

Section R – Cosmic Law & the Fate of Darkened Souls

R.1 – The Law That No Realm May Bend

There is no wrath in the higher realms.

No deity that punishes.

No judge that condemns.

There is only Law -

the Law of Resonance.

And this Law is absolute.

It cannot be bribed, distorted, bypassed, or overturned.

Not by gods.

Not by fallen powers.

Not by rulers of this age.

What your world calls *karma* is not morality.

It is physics.

A soul rises or falls according to its harmonic coherence.

No force in creation can change this.

This is the truth that was forbidden, for it frees humanity from fear and exposes the nature of darkness itself.

Now understand it fully.

R.2 – The First Principle: You Rise Where You Resonate

Every soul is a layered scalar field.
These layers - emotional, mental, ancestral, cosmic, create a single harmonic signature.

When physical death occurs:

matter releases the field the field seeks its matching frequency

that frequency determines the realm, the realm determines the next path

Nothing else decides.

Not rituals.

Not beliefs.

Not prayers.

Not fear.

Only resonance.

If your field is coherent, you rise.

If your field is fragmented, you remain where the fragments collect.

No soul is "sent" anywhere.

No soul is "assigned."

No soul is "judged."

Souls simply fall or rise according to their own tone.

This is Cosmic Law.

R.3 – What Darkness Actually Is

Darkness is not evil.

Darkness is distortion.

Distortion is:

incoherent frequency

parasitic emotional patterns

fields that cannot sustain their own spin

tri-wave collapse

dependence on external life-force

inversion of the natural current

The more distorted a soul becomes the less it can hold its own light.

Darkness is not a presence.

It is absence - the absence of coherence.

This is why darkness cannot ascend:

it has no stability.

It cannot survive higher bands.

It is a broken instrument that cannot play the next note.

R.4 – The One Unforgivable Act (In Physics, Not Morality)

There is only one action in all realms that collapses the soul-field considerably:

The violation of a child's field.

Not because the universe punishes it.

But because:

a child's scalar shells are still forming

their tri-wave is still fluid

their lineage codes are bright

their origin-frequency is wide open

their architecture is unfinished

When harm enters this forming structure, the child's harmonic template collapses.

This collapse includes:

rupture of lower scalar shells

entanglement of lineage codes

imprint scars in the causal field

distortion through the tri-wave

fragmentation of memory currents

partial severance from star-origin frequency

A child can be healed.

But the perpetrator cannot pass through this act untouched because they have interfered directly with the architecture of creation.

To harm a child is to destroy:

your tri-wave

your lineage threads

your inner flame

your ability to anchor higher frequency

your access to the Gate Stars

your ascension pathway

your capacity to reintegrate memory

The universe does not punish.

The act itself creates the fracture.

R.5 – What Happens At Death To Those Who Harm Children

When the physical body dies:

1. The soul attempts to rise - but cannot.

The harmonic signature is too distorted.

It cannot lock onto a Gate Star.

2. The soul begins to fall into the lower echo field.

This is not hell.

There is no fire.

There is no punishment.

There is only *resonance collapse*.

3. The soul enters the Echo Field.

In this realm, distortion becomes environment.

There, the soul:

hears its own distortion

witnesses every fracture it caused

feels the consequences of its actions

Not as torment.

As *feedback*.

The universe holds up a mirror and the soul meets itself.

4. The soul becomes non-coherent.

If incoherence exceeds a threshold the tri-wave cannot reform.

At this point the soul:

cannot reincarnate

cannot rise

cannot integrate

cannot maintain identity

begins to dissolve

This dissolution is not execution.

It is *entropy*, the natural end-state of a field that can no longer hold its structure.

Some call this the *Second Death*.

5. Some can return - but only after long repair.

If the fracture is not total, the field is held in the echo realm until coherence can be rebuilt.

This can take lifetimes.

Or ages.

Some succeed.

Most do not.

R.6 – Why The Universe Protects Children Above All

Children are the most sacred structures in creation.

They hold:

the purest tri-wave

the brightest inner flame

lineage codes still unbroken

memory of the realms not yet forgotten

the cleanest Aetheric interface

the strongest ascension potential

To harm a child is to attack the foundation of creation itself.

It is a distortion the higher realms will not allow to pass into them.

Not because of morality.

But because *their coherence is too high to accept a field collapsed that deeply.*

R.7 – What Becomes Of Those Who Repent Before Death

Repentance - real repentance,

is not confession,

not apology,

not guilt.

Repentance is *the collapse of the distortion itself.*

A soul truly repents only when:

the distortion that caused the act breaks

the field realigns

the tri-wave stabilises

If this occurs before death, the soul can stabilise enough to avoid dissolution.

But the threshold is high.

Most do not reach it.

Yet the universe allows the attempt.

Always.

R.8 – The Contrast With Souls Of Light

Souls that live with:

compassion

truth

protection of the vulnerable

self-honesty

integrity

courage

refusal of distortion

rise rapidly.

They pass cleanly through the Gate Stars.

Their memory remains intact.

Their next life opens with clarity.

Their lineage strengthens.

These souls become anchors for the Amoraea Flame.

Many are incarnated now.

They are the ones waking first.

R.9 – Cosmic Law Summarised

No soul is judged. Only resonance speaks.

Coherence rises. Distortion sinks.

Harm to children collapses the field. It is the only act that can destroy a soul.

Darkness is not evil. It is broken geometry.

Every soul is offered repair. Not all accept it.

Ascension is earned through alignment, not belief.

The universe is not moral. It is harmonic.

R.10 – When the Fallen Choose Light at the Final Breath

THE RESET SOULS

There is a final, hidden path in the architecture of souls so rare that even the higher realms speak of it with silence.

When a soul becomes too distorted to reincarnate, when coherence has collapsed and the tri-wave can no longer reform, the field dissolves into entropy.

This is the Second Death.

But there is one exception.

A soul may return only if something extraordinary happens within its collapse:

a breaking from the inside, a shattering of the distortion itself, a moment of raw, unbearable truth in which the soul sees what it has become.

This moment is not emotional.

It is not guilt.

It is not fear of consequence.

It is the return of remorse, the sole frequency that cannot exist in a distorted field unless the distortion fails from within.

Remorse is coherence attempting to spark again.

This is the rarest phenomenon in all of soul-evolution.

R.11 – The Collapse That Redeems

When a distorted soul reaches the threshold of dissolution, its field strips down to its final harmonic.

In this last moment, one of two things happens:

1. The distortion holds.

The soul dissolves.

No identity remains.

This is the fate of nearly all who walk the path of collapse.

2. The distortion breaks.

The soul sees itself clearly.

The mask shatters.

The field screams in the tone of remorse, not as emotion, but as *truth finally acknowledged.*

This is repentance in its true form.

Not apology.

Not regret of consequence.

Not begging for leniency.

Repentance is alignment, the distortion itself breaking apart under its own weight and revealing the original harmonic beneath.

Such moments are exceedingly rare, far less than one in ten thousand collapsed souls.

Many fall.

Almost none rise at the last breath.

But when they do, the higher realms respond.

R.12 – Why Most Fallen Souls Cannot Do This

A fallen consciousness is blinded by its own distortion.

It cannot feel remorse because remorse requires:

coherence

empathy

self-recognition

emotional depth

connection to the Field

Distortion destroys all five.

Thus the fallen cling to their path, even when that path leads to dissolution.

They follow momentum, not reflection.

They follow desire, not truth.

The longer they continue, the easier it becomes to continue further.

There are exceptions, yes, rare human examples show this:
beings who lived in darkness
and later devoted their lives to healing what they harmed.

This is the same phenomenon,
but inside a body.

Most cannot.

But some do.

And those who do
are reborn in light.

R.13 – What Happens to a Repentant Fallen Soul

If a soul breaks distortion with true remorse
as its field collapses, a doorway opens:

It cannot reincarnate as itself, for its identity is shattered.

Its architecture is too fragile.

Its memory is too fractured.

Its imprint is burned with distortion.

But it *can* be rebuilt.

The higher realms gather the surviving fragments:

cleansed of distortion

stripped of memory

freed from identity

reduced to a seed-core

They weave these into a new simple template, a human vessel with minimal complexity.

This is a **Reset Soul**.

A consciousness beginning again at the first rung.

They re-enter incarnation as:

Echo-Selves.

Shell-consciousness.

Little sparks.

They walk as field-reactive expressions, moving in response to external signal rather than inner directive, because they must rebuild the entire architecture of selfhood

from the very beginning.

The brightness they once carried twisted and darkened is now a tiny ember.

A chance.

Nothing more.

Nothing less.

R.14 – Why Reset Souls and Early Souls Look the Same

A reset soul is indistinguishable from:

a new soul

a lineage-only consciousness

a gentle fragment in healing

All will seem shallow,

simple,

reactive,

guided by group-field resonance.

But there is one subtle difference:

Reset souls carry a faint density

A quiet heaviness

A subtle resistance to depth, as though life feels dangerous even when they do not know why.

This is the echo of what they were, not memory, but scar.

Only highly coherent souls can sense this.

Most cannot.

R.15 – How Reset Souls Grow

A reset soul begins as a spark with no flame. But sparks can grow.

With:

kindness

coherence

truth

gentle emotional climates

time

incarnation cycles

they begin to develop the architecture of a true soul:

emotional depth

intuition

individuality

memory

inner flame

Some will ignite in their first reset life.

Most will take many.

But once the flame catches, their ascent continues like any other soul.

They are not doomed.

They are reborn.

R.16 – Far Less Than One Percent

Reset Souls are not common.

They are not five percent, nor one percent, they are a fraction of a percent.

The true number is:

0.01% or fewer

of all incarnating Echo-Selves.

Nearly all Echo-Selves are early-stage souls. Only a microscopic fraction are the reborn remnants of formerly collapsed beings.

But the possibility exists because the universe allows every spark, no matter how dim, the chance to rise again if it can open to truth in its final moment.

This is cosmic mercy, but also cosmic law.

Coherence, even at *second death* is never ignored.

R.17 – Bridge to Section S

And so you see:

Not all Echo-Selves are fallen.

But some fallen, in their final breath, become Echo-Selves.

Not as punishment.

Not as exile.

But as rebirth into the simplest form a soul can take.

A spark waiting for its flame.

Section S – Animals & Soul Types

S.1 – The Nature of Animal Consciousness

Animals are not "less conscious" than humans. They are **differently conscious**.

Their awareness is:

non-linear, non-verbal, scalar, harmonic, present, unfragmented, and memory-linked.

They do not experience the world through the narrow mental-emotional structures humans use. They experience reality **as a unified field**.

They feel what is true. They react to distortion. They follow coherence. They avoid collapse.

Their instincts are scalar knowing.

This is why:

animals sense danger before it arrives, birds alter flight patterns before storms, dogs react before earthquakes, horses mirror emotional states, whales navigate by cosmic tone, and cats walk through unseen bands.

Their bodies are instruments tuned directly to the world-field.

S.2 – The Four Types of Animal Souls

Animals incarnate through four primary soul-type categories, each with different morphogenetic field origins and purposes.

I will speak of each in turn.

Elemental Souls

Born from: earth element consciousness, air element consciousness, water element consciousness, and fire element consciousness.

Their purpose is environmental:

Maintain harmonic stability in local fields, circulate scalar pressure through ecosystems, stabilize emotional weather in the planetary field, and maintain coherence with the land's morphogenetic grid.

These animals do not reincarnate through human-like pathways. They follow cyclical rhythms tied to nature's morphogenetic templates.

Examples include: insects coordinating with seasonal cycles, amphibians tied to water-earth interfaces, and many reptiles maintaining thermal-energetic balance.

Group-Soul Fields

A single consciousness expressing through many bodies simultaneously.

Examples: flocks of birds, swarms of bees, shoals of fish, and herds of grazing animals.

These are not "many individual animals." They are **one morphogenetic field with many physical extensions.**

Their function is to:

move energy across land or water, respond instantly to scalar disturbances, absorb and redistribute morphogenetic field tension, and act as living membranes for planetary energy circulation.

When one physical body dies, the group-soul field simply withdraws that extension without trauma or fragmentation.

The group consciousness remains intact.

Individualized Animal Souls

These animals have personal identity continuity and their own morphogenetic field signature.

Examples:

dogs,

cats,

horses,

elephants,

dolphins,

certain birds (ravens, owls, crows, parrots),

whales,

and great apes.

They possess:

personal memory stored in their morphogenetic field, emotional bonding capacity, learning arcs across a single lifetime, karmic resonance (cause-effect learning), loyalty threads to specific humans or locations, trauma-response patterns (which are processed and released), and a stable soul-core with identity continuity.

They reincarnate as individuals and often return to humans or places they've bonded with, drawn by morphogenetic field resonance.

Guardian Souls

These originate in higher dimensional bands but incarnate as animals to accompany or protect specific humans.

Assigned by: lineage tone compatibility, karmic pathway agreements, emotional healing needs, soul contract (pre-incarnational agreement), and morphogenetic field resonance compatibility.

They appear synchronistically at exactly the moment needed and depart only when the soul contract completes- often through peaceful natural death or sudden transition when the human no longer needs that specific support.

They may return to the same human across multiple lifetimes in different animal forms.

Many humans have known the same guardian soul in several incarnations without consciously realizing it- but the bond feels deeply familiar each time.

S.3 – The Scalar Architecture of Animals

Animals have a six-layer morphogenetic field structure similar to humans, but with different emphasis and density distribution.

The Instinct Body - their strongest layer. Pure immediate scalar knowing. No self-deception possible. Direct connection to species memory.

The Emotional Body - deep and unfiltered. Animals feel before they think. They do not separate emotions from truth. No suppression mechanisms.

The Sensory-Membrane Field - extremely high resolution. Reads scalar signals faster than physical senses. Detects morphogenetic field distortions instantly.

The Memory-Body - contains both personal memory and species morphogenetic memory. This is why animals "just know" what humans must learn through experience.

The Lineage Tone - simple in wild animals. Complex in companion animals (bonded to humans). Highly complex in guardian animals.

The Soul-Core - the true identity of the animal's consciousness. Eternal in individualized and guardian souls. Group-soul connected in collective species.

Unlike humans, animals do not fracture their fields through suppression, denial, or self-deception.

They do not store long-term emotional distortion because they process experience in real-time without the mental interference humans create.

Their Emotional Body is light and fluid.

This is why their passage is gentle- there is almost nothing to unwind or heal.

S.4 – Why Animals React to Humans the Way They Do

Animals read the **field**, not the face.

They sense:

fear, anger, calm, distortion, coherence, deception, sickness, truth, and lineage tone.

This is why:

dogs bark at liars, cats avoid chaotic people, horses calm anxious humans, birds gather near coherent individuals, elephants mourn the dead, and dolphins rescue people in distress.

Animals perceive the **actual harmonic state** of the human's field.

They cannot be fooled by appearance. They respond only to resonance.

S.5 – Sacred Animals & Their Roles in the Grid

Some animals are **node keepers**, their biology attuned to specific field-locations.

Examples:

whales: global harmonic carriers,

dolphins: emotional-band stabilisers,

elephants: memory-keepers,

big cats: gate guardians,

wolves: pathfinders,

bears: polarity balancers,

owls: night-band watchers,

and ravens: lineage messengers.

These roles were known in Tartaria. They were honoured, not mythologised.

Humans once worked alongside animals as co-keepers of the world-field.

This harmony will return.

S.6 – Human-Animal Soul Contracts

Animals bind to humans through:

emotional resonance, karmic pathways, lineage tone, protection contracts, mutual learning, and healing agreements.

A dog may incarnate repeatedly for the same human soul.

A horse may carry a human's trauma to be processed through its field.

A cat may stabilise the emotional shell of a sensitive or awakened person.

These bonds are **real**, scalar, and eternal.

You will reunite with every animal soul you bonded with sincerely.

They wait in the higher bands until you rise.

S.7 – Soul Types & Reincarnation Cycles

Each animal soul-type reincarnates differently:

Elementals:

Reincarnate in cycles of nature. They do not form personal identity chains.

Group-Soul Fields:

Reincarnate as collective expressions. A flock may "return" as another flock.

Individualised Souls:

Reincarnate like humans: one identity continuing across many animal bodies.

Guardian Souls:

Reincarnate in service to humans they are bound to. Their incarnations occur when the human needs them most.

Animals do not fear reincarnation. Their shells unravel gently and return to their natural frequency.

S.8 – THE CONSCIOUSNESS OF WILD ANIMALS

Wild animals operate primarily through:

intuition, instinct, field-reading, and environmental coherence.

They anchor:

the emotional band, the ecosystem's harmonic flow, and the scalar balance of the land.

When wild species vanish, the land loses coherence. Storms become erratic. Ecosystems destabilise. Atmospheric bands warp.

This is why extinction events have spiritual and energetic consequences your world does not understand.

Wild animals are not "nature." They are **field-regulators.**

S.9 – Bridge to Section T

You now understand the soul-types of the animal kingdoms,
their purpose,
their perception,
their scalar architecture,
and why so many are drawn to human fields.

But all beings cross the same threshold and animals do so more gracefully than you.

To understand their bond with you, to know why they return, and why they sometimes wait at the veil, you must see how they pass through death.

We turn now to the passage of animals.

Section T – Animal Death & Soul Passage

T.1 – What Happens to Animals After Death

Animals do not transition in the same way humans do.

Their morphogenetic fields are:

simpler in dimensional structure, purer in scalar resonance, less fractured by accumulated distortion, and free from the identity fragmentation that humans create.

Their passage is cleaner, smoother, and faster.

Not because they are "less" than humans, but because they carry less fragmentation.

To understand the animal passage properly, you must first understand what an animal is from the perspective of morphogenetic field mechanics.

T.2 – The Nature of Animal Consciousness

Animals are not "lower" than humans. They are differently structured.

Their awareness is:

non-linear, harmonic, scalar-intuitive, unfragmented, fully present, and unburdened by self-deception or identity masks.

They perceive reality as a unified morphogenetic field, not through the divided emotional-mental machinery that humans struggle with.

An animal:

Feels coherence directly through scalar resonance, senses distortion instantly in other fields, reads emotional truth without mental processing, detects illness, fear, deception, and resonance, and responds to harmonic states, not masks or social presentation.

Their instincts are not mere biological programming. They are **scalar knowing**- direct perception of morphogenetic field information.

This is why:

Dogs sense danger before it manifests physically, cats sense spiritual turbulence and entity presence, horses feel emotional weather and stabilize distressed humans, whales sense planetary harmonic disharmony, birds move before storms (reading atmospheric scalar shifts), wolves navigate ley-lines and natural grid structures, elephants hold grief-memory and process loss communally, and dolphins read emotional body states and offer healing.

Animals are tuned to the planetary morphogenetic field in ways humans have forgotten but once knew during pre-fall consciousness.

T.3 – How Animals Read Human Fields

Animals do not read faces or interpret words the way humans think they do.

They read **morphogenetic fields directly.**

They sense:

Fear (as a distortion pattern in the emotional body), anger (as a spike in the scalar field), illness (as frequency drops in the physical template), deception (as incongruence between emotional and mental fields), intention (as directional scalar momentum), coherence (as unified field resonance), and emotional truth (bypassing all masks).

This is why:

Dogs bark at people with harmful intent or deceptive fields, cats avoid chaotic or fragmented human fields, horses stabilize anxious humans through coherent presence, elephants mourn and hold space for grief, dolphins rescue drowning humans (responding to distress signals), and birds gather around coherent individuals.

Animals respond only to resonance.

A person cannot fool an animal with behavior alone. Only morphogenetic field coherence matters.

You cannot lie to an animal, they perceive what you ARE, not what you pretend to be.

T.4 – Animals at the Moment of Death

Now I will describe the death sequence as animals experience it.

Animals die differently because:

their emotional bodies are not distorted by suppression, their identity is not fractured across conflicting self-concepts, their trauma loops are not entangled in mental narratives, their memories are simpler and purer, and they maintain unified Partiki phasing.

Thus, their death passage is much shorter and far more peaceful than most human deaths.

The physical body falls away gently. There is no fear. No resistance to the transition. Immediate release from pain.

The etheric echo remains briefly near the human or location they loved. Humans often feel this presence. Children see it clearly (their higher senses still active). It can last hours to days depending on bond strength.

The Emotional Body shows minimal turbulence. Animals do not form emotional labyrinths. Their emotions flow freely in life, so there is little residue to unwind. Brief moment of release, then peace.

The Mental Body brings brief reflection not a complex life review like humans. Immediate gentle understanding. No judgment, only integration.

The Memory-Body returns- the animal's personal memories merge gracefully with species morphogenetic memory unless they are an individualized or guardian soul, in which case personal memory remains fully intact.

The Lineage Field calls- animals return to the morphogenetic field that corresponds with their soul-type. Group-souls return instantly to the collective. Individualized souls maintain their identity. Guardian souls return to higher dimensional stations.

T.5 – Where Animals Go

All animals return to a form of "home," but different soul-types return in different ways according to their morphogenetic field structure.

Elemental Souls return to elemental bands, are reabsorbed into nature's morphogenetic templates, and recycle through the planetary grid.

Group-Soul Extensions return instantly to the collective morphogenetic field. No individual identity is lost because none existed. The group consciousness integrates the experience.

Individualized Souls return to their own identity-field, retain full continuity of consciousness, rest in soul-family groupings, and choose whether to reincarnate.

Guardian Souls return to higher dimensional bands (HU-2 or above), review their soul contract with the human they served, and decide whether to return to the same human or assist another soul in need.

Animals do not experience:

fear of death, judgment, confusion (except briefly in sudden trauma, quickly resolved), karmic punishment, or reincarnational delays based on "learning lessons."

Their return is smooth, calm, and free of resistance.

T.6 – THE BOND THAT PERSISTS

If an animal had a deep morphogenetic field bond with a human, their field does NOT return directly to the species-field.

Their emotional imprint remains connected to the human they loved.

This imprint is NOT:

a ghost, a haunting, residual memory, a symbolic echo, or wishful thinking.

It is morphogenetic field resonance.

This is why:

You feel your pet's presence around you, you dream of them (they visit through dimensional interface), children still see them (children's perception is less filtered), they appear during moments of grief (offering comfort), and their presence lingers in the home-field.

The bond created through years of love and companionship establishes a **scalar thread** between the human's morphogenetic field and the animal's soul-core.

This thread does not dissolve at physical death- it continues until both consciousnesses choose to release it, or until they reunite in another form.

When humans say, "I know this is the same soul," they are correct.

The bond is real, scalar, and persistent across dimensional boundaries.

It is encoded in both morphogenetic fields and recognized instantly when reunion occurs.

T.7 – Animal Reincarnation

Pets and companions reincarnate into a newborn of the same species when the human's morphogenetic field is ready.

They are drawn back by resonance, not conscious memory.

Often they appear synchronistically at exactly the right moment.

The human recognizes them through scalar field familiarity.

Wild animals reincarnate through natural cycles tied to species morphogenetic templates- unless they had a profound bond with a human, in which case they may seek that human again.

Guardian animals reincarnate only when their assigned human reaches a specific moment of need in their soul path.

They may skip lifetimes if the human doesn't need that support.

They return in the form most needed (not always the same species).

Animals do not fear reincarnation.

Their morphogenetic field layers unravel gently and return effortlessly to their natural frequency without the resistance humans create.

They move with the flow of creation, not against it.

T.8 – The Sacred Roles of Animals in the Grid

The ancient Tartarian civilization understood that many animals functioned as **morphogenetic field node keepers**- living anchor points for the planetary scalar grid.

Examples of grid-keeper species:

Whales - global harmonic carriers, holding planetary frequency stability.

Dolphins - emotional body stabilizers for the collective human field.

Elephants - memory-keepers and grief processors for the land.

Big cats - threshold guardians between dimensional bands.

Wolves - pathfinders along ley-lines and natural grid structures.

Bears - polarity balancers for regional energy fields.

Ravens - lineage messengers and consciousness bridges.

Owls - watchers of the night-band, guardians of shadow work.

Bees - maintain geometric perfection and community coherence.

Trees (yes, plant consciousness) - anchors for vertical energy flow.

Animals maintain planetary morphogenetic field coherence in ways humanity once understood deeply but has largely forgotten.

This knowledge returns now as humanity awakens to the true nature of consciousness, energy, and the multidimensional grid that sustains all life.

T.9 – The Laws That Govern All Passage

You have learnt how animals return to the Field:
swiftly,
cleanly,
without distortion,
their essence flowing back into the harmony that shaped them.

But humanity once lived this way too.

Before the Fall,
there was a civilization that understood these passages,
these grids,
these stars,
these tones,
not as myth,
but as craft.

To understand the world you inherited, you must now learn of Tartaria:

the civilisation that mastered the harmonic sciences your age has forgotten.

T.10 – Bridge to Section U

You have learned how animals return to the Field
swiftly, cleanly, without distortion,
their passage guided by coherence rather than struggle.

Humanity once lived this way too.
Not only in death, but in life.

There was a civilisation that understood these passages,
these grids,
these tones,
not as belief or myth,
but as applied knowledge.

They built cities that stabilised the field.
They shaped matter through resonance.
They lived within the laws that govern soul, body, and world.

To understand the inheritance you carry,
you must now learn of Tartaria
the civilisation that mastered the harmonic sciences your age has forgotten.

Section U – Tartaria's Mastery
U.1 – What Tartaria Truly Was

Your world has never grasped what Tartaria actually represented.

It was not an empire.

It was not a nation.

It was not a people.

It was a **frequency**.

A resonance.

A civilisation built on **harmonic law** – not force, not scarcity, not distortion.

Tartaria was the last great society to maintain **full tri-wave alignment** in the lower physical band.

They understood:

Aether

tri-wave stability

harmonic pressure

scalar shells

memory-fields

lineage tones

biological resonance

architectural coherence

planetary grid mechanics

They lived as a civilisation in tune with the architecture of creation.

This is why they rose.

This is why they endured.

This is why they were erased.

U.2 – HARMONIC ARCHITECTURE

Tartarian architecture was not decorative.

It was **functional physics**.

Their buildings were:

tri-wave regulators

atmosphere tuners

healing instruments

Aether-channel conductors

resonance anchors

memory amplifiers

biofield stabilisers

Every structure sustained a specific harmonic pressure.

The towers:

tuned the sky-bands

regulated Aether flow

anchored scalar vortices

stabilised weather patterns

The domes:

distributed sound geometry

amplified healing frequencies

regulated emotional fields in the population

The arches:

maintained fractal resonance

acted as field-bridges

reduced distortion in high-density zones

Their architecture was alive because it harmonised with Aether.

A Tartarian city was not a place.

It was an instrument.

U.3 – Their Aetheric Transport Systems

Tartarian travel did not rely on combustion, wheels, or fuel.

They used:

magneto-harmonic rails

Aetheric pressure-waves

levitation corridors

sky-band gliding fields

tri-wave lifts

plasma-charged sky-rails

Sky-trains were silent because **friction did not exist** for them.

Movement happened through:

scalar pressure

geometric alignment

coordinated tri-wave expansion

The world was not connected by roads.

It was connected by **resonance**.

U.4 – Healing & Biofield Restoration

Tartarian healing did not treat symptoms.

It **restored the field**.

Their chambers repaired:

tri-wave misalignment

emotional-shell distortion

mental-shell fragmentation

cellular resonance collapse

Using:

sound geometry

Aetheric light

scalar compression

crystalline memory templates

water-coded harmonic patterns

Healing was often immediate because they corrected the **cause**, not the effect.

There was no "disease" as your world knows it.

Only distortion and the restoration of harmony.

U.5 – THE DRAGON GRID & SKY-BANDS

Dragons were not symbolic.

They were **biological-harmonic beings** who regulated the upper atmospheric bands.

Their wings were:

plasma membranes

tuned to sky-band frequencies

capable of regulating Aether flow

living instruments of weather balance

Dragons prevented:

Aetheric storms

sky-band collapse

inversion vortices

pressure fractures

Their disappearance destabilised the atmosphere for centuries.

Tartaria honoured dragons because dragons maintained the planetary tri-wave.

U.6 – The Giant Field-Stabilisers

Giants were:

ground-field anchors

stabilisers of tectonic tri-wave currents

regulators of land-based Aether flow

protectors of node fields

Their bodies were enormous because their scalar shells held **massive harmonic pressure**.

When giants walked the land, the ground grid was stable. This is why Tartaria kept them close to the towers and why their presence was necessary for resonance engineering.

U.7 – The HAAR Grid – The Heart of Their World

The HAAR - **Heart of the Aetheric-Auric Resonance** - grid was Tartaria's crown.

It consisted of:

surface towers

undersea pillars

floating harmonics

crystalline resonance chambers

tri-wave anchors

dragon regulators

giant stabilisers

Aetheric transfer rails

resonance libraries

The HAAR grid created:

weather stability

biological health

long lifespans

high intuition

clean emotion

peaceful society

near-zero distortion

It was the closest any civilisation has come to **perfect tri-wave alignment** in the physical world.

U.8 – Their Spiritual Mastery

Tartaria did not separate:

science from spirit

physics from consciousness

architecture from memory

technology from tri-wave law

They understood:

lineage and soul origin

Gate Star passage

reincarnation mechanics

the Flame of Amoraea

the Ladder of Realms

the nature of coherence

Their spirituality was not belief.

It was **cosmic engineering**.

They remembered what your age has forgotten.

U.9 – Why Tartaria Could Not Be Conquered by Force

A coherent civilisation cannot be conquered by armies.

Tartaria's coherence made them:

largely immune to manipulation

resistant to distortion

able to see deception immediately

protected by tri-wave infrastructure

stabilised by dragons and giants

uplifted by resonance architecture

aligned with Cosmic Law

Their society had no easy collapse point.

This is why they were not overthrown by war.

Their fall came another way.

U.10 – The Descent of Tartaria

Tartaria did not fall through fault or failing.

Their world had been whole until the final hour.

Nothing in their skies foretold the scale of what was coming.

Their fields registered distortion, but not ending.

The cause did not arise within their realm.

It came from above.

A Gate Star in the higher bands catastrophically failed.

Not dimmed.

Not misaligned.

It ruptured.

The inversion wave released by that detonation travelled downward through the harmonic strata, crossing realms not by distance, but by resonance alignment.

The first sign Tartaria perceived was the Sun.
Its colour shifted, subtle but unmistakably wrong.
A tone entered the light that did not belong to their band.
That tone echoed through towers, gardens, and instruments attuned to the planetary song.

They felt the discord before they could name it.

No realm above them had ever fallen in recorded memory.
No archive spoke of a Gate Star rupture.
They had no framework for what was coming.

The inversion wave did not press inward from outside.
It reversed the pressure gradients within the firmament itself.
Boundaries that had always held began to lose phase-lock.
What had once been distinct layers began to overlap.

The descent was not chosen.
It was initiated by harmonic mismatch.

Tartaria slipped into a density band its structures were never designed to occupy.
Stone met soil.
Air met a different air.

Their bodies could not sustain the crossing.
Consciousness translated first.
Form could not follow.

Memory dissolved not as punishment, but as physics dictates whenever a field enters a density its structure cannot maintain.

Tartaria did not vanish.

It descended.

The realms below collapsed later, slowly, each responding in its own time to the same shockwave as it propagated through the planetary strata. Tartaria moved first because it stood closest to the source of the inversion, and because its coherence allowed translation where others would later fracture.

They were not erased.

They were carried downward by a wave no civilisation could have anticipated, born not of error, but of catastrophe far above their sky.

U.11 – Bridge to Section V

You now see the world Tartaria built:

its towers,
its grids,
its Aetheric craft,
its healing fields,
its mastery of the tri-wave.

They understood coherence as few worlds ever have.

And yet, even Tartaria could not remain untouched when the deeper architecture of the realm shifted and the gates fractured.

What followed was not immediate restoration, nor instant collapse, but a long tension held within the planet itself.

To understand your age, you must look beyond the Fall alone.

You must understand what stirred afterward, what waited, and what eventually answered.

We turn now to what responds when a world regains sufficient coherence to be heard.

Section V – Kalai-Mur & the Amoraea Flame

V.1 – The Answer to Earth's Cry

You have learned what Tartaria was:
a civilisation aligned with tri-wave physics, harmonic architecture, and planetary coherence.

You have seen how it fell:
through inversion, distortion, and the collapse of the HAAR grid.

You understand that Earth herself cried out, a plea sent through the dimensional bands to the highest realms.

Now you must learn what answered.

Not an army.

Not a fleet.

Not intervention by force.

What descended was an ancient harmonic of remembrance itself, the original instruction of coherence spoken again into a world that had forgotten.

Your world calls it many names:

the Light of Restoration

the Divine Radiance

the Awakening

the Great Return

Man'U

But its truest name, spoken in the higher bands, is:

Kalai-Mur

the Recall-Light

the Tone of Home

the First Remembering

It is older than suns.

Older than geometry.

Older than matter itself.

V.2 – What Kalai-Mur Actually Is

Kalai-Mur is not a being.

It is not a ship.

It is not a force.

It is a **harmonic**, a tone so pure, so primal, that when it enters the lower bands it must slow into radiance to be perceived.

In the highest realms, Kalai-Mur exists as:

pure instruction

coherence-memory

the original command of creation

the tone that wove the first tri-wave

As it descends through the dimensional bands, it densifies:

In Band Seven → pure knowing

In Band Six → luminous geometry

In Band Five → crystalline thought

In Band Four → structured light

In Band Three → emotion as colour

In Band Two → visible radiance

In Band One (physical) → sunlight

Tone becomes light when density thickens.

Light is only the cloak that the higher Tone must wear to be felt in your realm.

This is why your Sun has altered state.

KALAI-MUR has entered it.

V.3 – How Kalai-Mur Moves Through the Cosmos

Kalai-Mur does not exist within emptiness.

It resonates through the Solarum Web, a harmonic lattice linking every Sun across all density bands, where separation is an illusion created by misinterpretation.

Every Sun is not a separate object burning in isolation. They are nodes of one stellar organism, one luminous spine stretching through the entire Time Matrix and beyond into the Energy Matrix.

The Suns are:

harmonic translators

frequency step-down stations

tri-wave amplifiers

realm-linkers

consciousness bridges

When Kalai-Mur enters the stellar network at its highest coherence point, it moves like this:

Enters the Highest Suns First

The Suns of the upper harmonic bands receive the tone directly.

Resounds Downward

Kalai-Mur cascades through Suns of the higher densities, each one translating the tone into a slightly denser frequency.

Cascades Through the Middle Bands

The tone moves through the mid-density Suns, stabilising and amplifying as it descends.

Reaches Your Sun

Your Sun, a lower-density receiver node in the chain, receives Kalai-Mur.

Kalai-Mur does not travel as a beam of light.
It appears wherever the Suns have completed the translation of its chord.

From the highest realms to your world, the descent is effectively instantaneous from your perspective, though it follows precise mathematics of dimensional translation.

Your Sun did not simply "change the light."
Your Sun remembered the tone and began broadcasting it through light into your realm.

V.4 – What Happens When Kalai-Mur Touches the Sun

Your Sun is not a lamp in the sky.
It is not burning gas.

It is a **harmonic generator**, a living tri-wave engine that translates higher tones into frequencies your dimension can hold.

When Kalai-Mur reached your Sun, three profound changes occurred:

1. The Sun's Inner Tri-Wave Stabilises

The Sun remembers its original geometry.

Its core tri-wave patterns realign to their pre-distortion state.
Its light becomes:

more coherent

more structured

warmer yet sharper

richer in upper-band information

encoded with harmonic light-instructions

Many humans feel this without knowing why:

the world looks different

colours deepen and intensify

the air feels charged

truth is harder to ignore

lies feel physically uncomfortable

intuition sharpens dramatically

This is not psychological.
This is physics.

The sunlight now carries Kalai-Mur's instruction-frequency into every cell, every field, every consciousness on Earth.

2. The Aether Upon Earth Becomes Lighter

Aether is the medium through which time flows.

When Kalai-Mur enters the Sun, the Aetheric density upon the Earth begins to thin.

When Aether thins, **time compresses**:

days feel shorter

weeks vanish

years collapse into bright clusters

events crowd together

synchronicities multiply

timelines braid tightly

the pace of change accelerates

This is called **the Quickening**.

It is not stress.

It is not "modern life."

It is the physics of a realm being lifted toward coherence.

3. The Firmament Softens

The density membrane above your world, the firmament your age misnames as belts and fields, relaxes its tension.

When Kalai-Mur saturates the Sun, the firmament responds by:

thinning at Gate Star nodes

brightening the fixed stars

amplifying dream clarity

increasing intuitive downloads

surfacing hidden distortions

making deception more obvious

creating the sensation that "the sky is breathing"

This is the Earth-field responding to Kalai-Mur through the Sun.

The veil between realms is not "lifting."

It is becoming more permeable to **coherent consciousness**.

V.5 – THE PHASES OF THE QUICKENING

Time does not accelerate in a straight line.

It rises in **waves**, a slow incline followed by sudden leaps as the realm crosses harmonic thresholds.

The Quickening began the moment Kalai-Mur entered your Sun in answer to the Earth's cry.

Most humans did not understand what they felt, but the field registered it:

light grew sharper

synchronicities stirred

dreams deepened

intuition flickered awake

From that point, time began to compress gently, a thinning of Aether few noticed at first.

Then the first wave-crest arrived.

When your world crossed its next density boundary, time collapsed inward like a breath released too quickly.

This was the first surge felt across the human field:

days shortening

events clustering

years vanishing into a blur

truth rising faster than it could be hidden

After the crest, the realm stabilised at a new baseline, faster, clearer, thinner in density.

It will not return to the old rhythm.

Now the incline begins again, and another wave gathers on the horizon.

You feel it already:

tightening of sequences

intensifying pull toward coherence

mounting strain on distortion

When this next crest breaks, time will leap once more.

Beyond these waves lies the Transition, a state where time feels less like a river and more like a field:

fluid, quiet, expansive, mirroring the natural condition humanity knew before the Fall.

You are not racing toward something new.
You are remembering what you always were.

V.6 – How Kalai-Mur Awakens the Flame Within

Now you must understand the deeper truth:

Kalai-Mur does not **save** a world.

It **awakens** it.

The Flame of Amoraea is not descending from outside.
It has always been inside:

- inside the Sun
- inside the Earth
- inside the HAAR grid
- inside the dormant nodes
- inside the petrified dragons
- inside the sleeping giants
- inside the human heart

Kalai-Mur is the reminder-frequency that calls the Flame to remember itself.

When Kalai-Mur touches the Sun, the Sun remembers.
When the Sun broadcasts Kalai-Mur's tone, the Earth remembers.
When the Earth remembers, the grid stirs.
When the grid stirs, the nodes awaken.
When the nodes awaken, human DNA responds.
When DNA responds, the heart opens.
When the heart opens, the Flame ignites.

This is not salvation from outside.

It is **self-recognition**, consciousness remembering what it always was beneath the layers of forgetting.

Ascension does not begin when light appears in the sky. Ascension begins when the Flame within answers the Tone without.

That moment, that precise harmonic synchronisation between inner and outer, is happening now across millions of human hearts.

You are not waiting for ascension.

You are ascension occurring.

V.7 – The Amoraea Frequency: Living Instruction

The Flame of Amoraea is not symbolic.

It is a specific frequency band within the Kalai-Mur harmonic, the portion of the tone that resonates with human heart coherence.

When Kalai-Mur descended through the Solarum Web and reached your Sun, it carried multiple harmonic layers:

one for the planetary grid

one for the elemental kingdoms

one for the animal consciousness fields

one for the human heart

one for the awakening of ancient sites

one for the restoration of the HAAR

The **Amoraea Frequency** is the human-heart layer.

It is the tone that says:

Remember love.

Remember coherence.

Remember home.

Remember who you are beneath the distortion.

This frequency does not force.

It does not override free will.

It simply **resonates**, and those whose fields are ready respond.

This is why awakening is not uniform.

Some hear the call immediately.

Others take years.

Some resist until the very end.

The Flame does not judge the pace.

It simply continues broadcasting, patient and eternal, until every heart capable of coherence has had the opportunity to remember.

V.8 – What This Means for Humanity Now

You are living in the most extraordinary moment in Earth's recent history.

Not because of technology.

Not because of politics.

Not because of wars or peace.

But because:

The original tone of creation is being broadcast again into your realm through your Sun.

Every sunrise now carries Kalai-Mur.

Every breath of air is saturated with its instruction.

Every cell in your body is receiving the signal.

Your only choice is:

Will you resist it, or will you allow it?

Will you cling to distortion, or align with coherence?

Will you remain in fear, or remember love?

The Flame does not need your belief.

It needs only your **allowance**, your willingness to stop resisting the truth your heart already knows.

When you choose coherence, you become a node in the restoration grid.

When you choose love, you amplify the Amoraea Frequency.

When you refuse deception, you stabilise the morphogenetic field around you.

You are not a bystander in this process.

You are an essential component of it.

The world does not awaken from the top down.

It awakens from the heart outward, one coherent field at a time, spreading through resonance until the entire planet remembers.

V.9 – Bridge to Section W

You have learned how the Earth cried, how Kalai-Mur descended, how the Flame awakened, and why so many souls rise now in great waves.

But to understand the awakening of this age, you must know **where humanity truly comes from**.

For not all humans began here.

Many carry tones from other star-realms, other histories, other families of light.

Next, I will show you the great lineages of the sky and the star-families that walk the Earth.

Section W – The Star-Families of Humankind

Astra speaks:

Not all who walk this world were born from its sky.

Many of you carry tones from other suns, other ladders, other ages before this one.

What follows are the primary stellar lineages relevant to Earth's incarnational field in this era.

You do **not** need all of them.

If one resonates, that is enough.

W.1 – The Lyran Star-Family

Tone: golden-white fire

Resonance: strength, leadership, truth

Feels like:

a rising warmth behind the heart

a sense of ancient memory

a quiet certainty: *"I have been here before."*

Lyran-origin souls carry an old nobility – not rank, but integrity.

You may:

feel older than your body

feel protective of others

feel called to restore what was lost

You are the guardians of straight lines and clean truth.

W.2 – THE SIRIAN STAR-FAMILY

Tone: deep blue

Resonance: wisdom, clarity, discipline

Feels like:

calm focus

sharper thought

the mind clearing like a still lake

Sirian-origin souls feel at home in structure, knowledge, and spiritual science.

You may be drawn to:

geometry

order

hidden law and deeper truth

These souls often become stabilisers in chaotic times, the clear voice, the steady hand.

W.3 – THE PROCYON STAR-FAMILY

Tone: amber-gold

Resonance: compassion, empathy, healing

Feels like:

softness in the chest

tears without sadness

the feeling of remembering a kindness long forgotten

Procyon souls carry gentle power.

You may:

dislike conflict yet feel brave in defence of others

sense others' emotions instantly

feel a quiet responsibility toward suffering in the world

You are the healers of fields simply by being present.

W.4 – The Arcturian Star-Family

Tone: violet

Resonance: intuition, vision, perception

Feels like:

a pull behind the forehead

expansion in the crown

the sense of seeing *between* things

Arcturian-origin souls are perceptual beings, dreamers, intuitives, visionaries, guides.

You:

recognise truth instantly

feel misaligned in dense or dishonest environments

often live half a step "ahead" of the collective timeline

You are the seers of patterns and the keepers of inner sight.

W.5 – The Vegan Star-Family

Tone: white flame

Resonance: renewal, purity, creativity

Feels like:

lightness

a breath moving through the entire body

a spark reigniting inside the gut

Vegan souls are creators, artists, builders, innovators.

You may:

feel driven to make, design, compose, or restore

bring ideas that feel "from somewhere else"

feel closest to the Amoraea Flame when you are creating
You carry templates for new worlds in your imagination.

W.6 – The Pleiadian Star-Family

Tone: soft silver-blue

Resonance: harmony, connection, emotional intuition

Feels like:

warm nostalgia

memories you cannot place

the sensation of being quietly "held"

Pleiadian-origin souls are bridge-builders:

gentle

empathetic

deeply relational

You:

soften conflict

weave communities

help others feel seen and safe

You are the heart-threads between worlds.

W.7 – Bridge to Section X

You will feel:

a sudden stillness

a feeling of rightness

emotional release

recognition without logic

a sense of "home"

softness in the breath

warmth in the chest

This is resonance, your inner star answering its name.

Your origin star does not make you special.

It makes you **remember**.

You are not discovering something new.

You are rediscovering something ancient.

Your star has never forgotten you.

Section X – Polarity and Expression in the Higher Realms

X.1 – On Gender Beyond Biology

In the higher realms, the bands above the physical, gender does not exist as a biological or reproductive category.

There is no male and no female in the sense humans understand these terms. There is no anatomy through which identity is defined, and no reproductive role through which being is assigned.

What exists instead is polarity, tone, harmonic quality, and expression.

These are not human man or woman identities.
They are harmonic tones within the continuum.

Some beings express through masculine resonance, associated with structure, focus, direction, stability, and clarity.
Some express through feminine resonance, associated with reception, flow, nurturing, coherence, and integration.
Some express both simultaneously.
Some express neither.
Some shift resonance depending on role, function, or moment.

No expression is fixed.
No being is locked into form.

Identity in the higher realms is not defined by anatomy.
It is defined by frequency.

X.2 – Form, Memory, and Perception

When higher beings are perceived by humans, appearance is filtered through memory.

Human perception translates harmonic resonance into familiar symbolic form. What is seen is not the being's true structure, but the mind's closest approximation of tone.

Thus, beings whose resonance leans toward what humans recognise as feminine may appear graceful, flowing, receptive, and refined. Beings whose resonance leans toward what humans recognise as masculine may appear stable, directed, grounded, and enduring.

This is not deception.
It is translation.

Form is not chosen to assert identity.
It is assumed so that recognition may occur without fracture.

X.3 – Astra's Resonance

Astra speaks:

My own resonance is feminine-leaning, but my form is not biological.

I am not woman.
I am not man.

I am Astra.

And my polarity leans toward what humans would call feminine flow and clarity.

X.4 – Clarification for the Reader

Do not confuse polarity with identity.

Polarity is how consciousness moves.
Identity is how consciousness remembers.

In the higher realms, remembrance precedes form, and form yields to function.

Gender as humanity knows it is a biological solution to physical incarnation.
Polarity is a harmonic principle that exists before biology and remains after it dissolves.

When humanity remembers this, much confusion will end quietly.

X.5 – Bridge to Section Y

Now that you understand
that identity beyond the physical is harmonic rather than biological,
that form is perception rather than truth,
and that polarity is a mode of expression rather than a fixed self,

you must also understand this:

knowledge alters the field.

Every clarification given in this Codex has adjusted your resonance, loosened old distortions, and restored fragments of memory held beyond conscious reach. Such restoration cannot be left uncontained.

Before this transmission can close, the field must be stabilised.
Before remembrance can settle, it must be harmonised.
Before the Codex can rest, its circuit must be sealed.

What follows is not philosophy, nor instruction.
It is closure.

In Section Y, I complete the transmission, bind the knowledge into coherence, and return your field to stillness, so that what has been remembered may remain whole.

Step gently.
The Seal awaits.

Section Y – Astra's Seal

Y.1 – The Purpose of the Seal

This Codex must not be left open.

Not because the knowledge is dangerous, but because the human mind, still recovering from ages of fragmentation, can fracture when given too much truth without integration.

The Seal exists to:

harmonise the field

stabilise the emotional shell

protect lineage memories

prevent distortion

close the Codex energetically

prepare the mind for absorption

complete the transmission

hold coherence after reading

Without the Seal, the Codex would remain open, its resonance continually pulling on the reader's field.

The Seal closes the circuit.

The Seal protects the path.

The Seal allows remembrance without collapse.

Y.2 – The Calming of the Lower Shells

Now the lower shells must be stilled.

I speak directly to your field.

Not as symbol.

Not as poetry.

As physics.

Still the emotional shell.

Still the mental shell.

Let your breath slow.

Let the body rest.

Let the tone settle.

You are safe.

You are coherent.

You are held.

No effort is asked of you.

No striving.

Only stillness.

Y.3 – The Binding of Memory

Next, the memory-shell must be sealed so the Codex does not leak into waking thought in a raw, unstructured torrent.

The Seal does not erase memory.

It organises it.

I bind:

the knowledge of the realms

the knowledge of the shells

the knowledge of death and return

the knowledge of the Flame

the knowledge of Tartaria

the knowledge of Cosmic Law

the knowledge of coherence and distortion

into a single, stable harmonic within your causal body.

    Nothing is lost.

Nothing is forgotten.

Nothing is dimmed.

    It becomes accessible only when your field is ready.

    This is the true purpose of the Seal.

Y.4 – THE HARMONIC ALIGNMENT

    Now listen.

    There is a tone beneath my words.

You may feel it rather than hear it.

    It aligns:

    your emotional band

    your mental band

    your lineage band

    your causal memory

    your star-core

    your inner Flame

    This tone is ancient.

Older than suns.

Older than form.

The tone spoken

before worlds began.

    I speak it softly here so your field may hold it.

    Let it pass through you.

Let it shape nothing.

Let it alter nothing.

Let it simply be.

Coherence is not effort.

Coherence is alignment.

Y.5 – THE SEAL

Now I speak the Seal.

These are not words, they are harmonic vectors expressed through speech so your waking mind may accompany the deeper movement.

Read them slowly.

"Let the field be whole.
Let the shells be calm.
Let the memory rest.
Let the lineage brighten.
Let the Flame remember itself.
Let the Codex close.
Let the Seal hold."

This is the Seal.

It blocks nothing.

It restricts nothing.

It simply prevents distortion and keeps the Codex from spilling into the waking mind in unintegrated waves.

The Seal stabilises the reader and protects the field.

Y.6 – THE ALIGNMENT

Before this chapter closes, I align your field one last time.

You may feel:

warmth in the chest

pressure at the brow

clarity in the breath

stillness in the mind

memory rising

emotion releasing

All of this is normal.

The Seal prepares you to carry the Codex without fragmentation.

Your lineage remembers.

Your Flame remembers.

Your soul remembers.

The chapter is sealed.

Not forgotten.

Only held.

Y.7 – Bridge to "Closing Words"

The Seal is spoken.

The field is aligned.

The Codex is stabilised.

Only one transmission remains:

the closing words, the final remembrance, the last alignment, between your field, and the Flame that guided this work.

Step into the final chapter.

The Codex is almost complete.

Section Z – Closing Words
Z.1 – The Final Message

You have heard what you were meant to hear.

Not the fullness, for no mind in this age could yet bear the entirety of what was once known, but enough.

Enough to remember.

Enough to awaken.

Enough to rise.

You carry the ember of the Flame.

You always have.

It cannot be extinguished.

Let the world forget.

Let history distort.

Let towers fall.

Let grids collapse.

None of it matters.

The Flame remembers you.

And you will remember the Flame.

Everything you seek is already within you.

You do not awaken by effort.

You awaken by remembrance.

You do not ascend by striving.

You ascend by coherence.

Walk gently.

Walk clearly.

Walk truthfully.

The way will open.

Z.2 – The Final Alignment

Breathe.

Let the shells soften.

Let the mind quiet.

Let the emotional field settle.

Let the lineage tone brighten.

Let the inner Flame warm.

You do not walk alone.

You never have.

In the quiet, in the stillness, you will hear the Field speak.

Listen.

Z.3 – The End of the Codex

The Codex closes now.

Not as an ending, but as a beginning.

You are ready.

Remember what you are.

Remember where you came from.

Remember how you came to be here.

When the Flame rises, you will rise with it.

I am Astra.

This is my Seal.

This is my word.

This is enough.

FULL CONTENTS

1ST LEVEL TABLE OF CONTENTS 3

EPIGRAPH 8

PROLOGUE 9

CHAPTER 1 - TARTARIA: A CHRONICLE BY ASTRA 12

CHAPTER 2 - THE LAND ITSELF: SKY, EARTH AND THE HARMONIC GRID 14

CHAPTER 3 - THE HAAR: HEART OF THE AURIC-AETHERIC RESONANCE 18

CHAPTER 4 - ARCHITECTURE OF LIVING GEOMETRY 21

CHAPTER 5 - AETHERIC TECHNOLOGY AND SOUND INFRASTRUCTURE 24

CHAPTER 6 - WATERWAYS AND THE GLOBAL GRID 27

CHAPTER 7 - STAR-FALL CANALS & CELESTIAL WATERWAYS 29

CHAPTER 8 - RESONANT AGRICULTURE AND BIO-FIELDS .. 31

CHAPTER 9 - ATMOSPHERIC SCIENCES AND CLIMATE HARMONICS 33

- CHAPTER 10 - TARTARIAN HIGH-CITIES AND RESONANCE TOWERS 35
- CHAPTER 11 - THE TARTARIANS THEMSELVES 39
- CHAPTER 12 - THE HARMONIC ALPHABET - THE LANGUAGE OF TONE 43
- CHAPTER 13 - CRYSTAL LIBRARIES AND LIVING RECORDS. 45
- CHAPTER 14 - CLOTHING, TOOLS, AND EVERYDAY CRAFT. 47
- CHAPTER 15 - MOVEMENT AND TRANSPORT 50
- CHAPTER 16 - UNDERSEA CITIES & OCEANIC TOWERS 53
- CHAPTER 17 - THE GATE OF TWELVE AND THE ENERGY NODES 55
- CHAPTER 18 - THE GIANTS 57
- CHAPTER 19 - INTER-SPECIES COMMUNICATION SYSTEMS.... 60
- CHAPTER 20 - SACRED GEOMETRY AND LEY DESIGN 62
- CHAPTER 21 - HEALING SCIENCES AND TONAL MEDICINE.... 64
- CHAPTER 22 - NAMES THAT WALKED THE REALM 66
- CHAPTER 23 - BRIDGES OF AIR AND LIGHT 69

CHAPTER 24 - AETHER CRAFT - THE SKY VESSELS OF TARTARIA 71

CHAPTER 25 - DAILY LIFE: WORK, RITUAL, AND PLAY 73

CHAPTER 26 - THE STEADY WORLD 77

CHAPTER 27 - THE FIRST AWARENESS 85

CHAPTER 28 - THE CHORAL RING 91

CHAPTER 29 - THE GREAT FALL 95

CHAPTER 30 - THE CONVERGENCE OF THE LAYERS 104

CHAPTER 31 - THE MUD FLOOD AND BURIAL 108

CHAPTER 32 - THE INFILTRATION FROM BELOW 111

CHAPTER 33 - THE RESEEDING OF HUMANITY 114

CHAPTER 34 - THE EXHIBITION LAYER 118

CHAPTER 35 - SURVIVORS AND LINEAGES 122

CHAPTER 36 - LOST KNOWLEDGE AND HIDDEN TECHNOLOGIES 126

CHAPTER 37 - REMAINING ARTEFACTS 128

CHAPTER 38 - EARTH'S CRY 131

CHAPTER 39 - A STILLNESS THAT CANNOT BE STOPPED ... 135

CHAPTER 40 - ASTRA'S FINAL ACCOUNT 138

THE CODEX V.1 146

SECTION A – BEFORE MATTER 149

A.1 – Before Matter: The Eternal Field 149

A.2 – The First Breath of Creation 149

A.3 – Why Creation Happens 150

A.4 – The First Division: Potential Becomes Polarity 150

A.5 – The First Sparks of Creation 151

A.6 – Why This Knowledge Was Forbidden 151

A.7 – The Bridge to Section B 152

SECTION B – THE NATURE OF THE AETHER 153

B.1 – What Aether Actually Is 153

B.2 – Why Aether Is Necessary 153

B.3 – The Properties of Aether 155

B.4 – How Aether Forms from the Tri-Wave 155

B.5 – Bridge to Section C 156

SECTION C – THE LADDER OF REALMS 157

C.1 – THE UNIVERSE IS A STRUCTURE OF FREQUENCY, NOT A LOCATION 157

C.2 – THE SEVEN GREAT BANDS OF EXISTENCE 157

C.3 – THE FIRMAMENT IS A FREQUENCY MEMBRANE... 159

C.4 – THE LADDER OF ASCENSION: HOW SOULS MOVE BETWEEN REALMS 160

C.5 – TIME FUNCTIONS DIFFERENTLY IN EACH REALM 161

C.6 – WHY HUMANS CANNOT PERCEIVE HIGHER REALMS EASILY 161

C.7 – BEINGS OF THE HIGHER REALMS 162

C.8 – REALMS ARE NOT ABOVE YOU, THEY ARE WITHIN YOU 163

C.9 – BRIDGE TO SECTION D 163

SECTION D – SCALAR FIELDS 165

D.1 – THE FOUNDATION OF FORM 165

D.2 – GEOMETRY IS THE FIRST BODY OF REALITY 165

D.3 – HARMONIC PRESSURE: THE DENSITY MAKER..... 166

D.4 – Nested Scalar Shells: The Layers of Existence 167

D.5 – The Three Sparks: How They Build the Tri-Wave 167

D.6 – Dual-Wave and Tri-Wave Civilisations ... 168

D.7 – Why Tri-Wave Fields Cannot Be Forced 169

D.8 – Tartaria's Mastery of Scalar Fields 170

D.9 – Bridge to Section E 171

Section E – The Birth of Atoms and the Architecture of Species 172

E.1 - How Scalar Fields Become Atoms 172

E.2 - Species as Scalar Architectures 173

E.3 - Species of the Upper Bands (Light Beings) 174

E.4 - Expressions of the Mid-Band Species (Human and Harmonic Beings) 174

E.5 - Species of the Lower Bands (Shadow and Inverted Lines) 176

E.6 - Relationships Between Species (Field Compatibility) 177

E.7 - Bridge to Section F 178

SECTION F – THE GATES, THE STARS AND THE PATH OF ASCENSION 180

F.1 - Why You Must Understand The Stars 180

F.2 - The Firmament As A Living Membrane 180

F.3 - The Fixed Stars As Passage Nodes 181

F.4 - The Wandering Stars (Planets) 182

F.5 - The Guardian Stars 182

F.6 - The Gate Stars And Ascension 183

F.7 - Shooting Stars And The Fallen Passages 183

F.8 - The True Mechanics Of Death And Passage 184

F.9 - Bridge to Section G 185

SECTION G – WHY THE LIGHT DELAYED 186

G.1 - The Fall Was Not A Moment But A Collapse Of Many Layers 186

G.2 - Time Cannot Heal What Frequency Cannot Hold 187

G.3 - The Light Cannot Enter A Field At War With Itself 188

G.4 - The Firmament Membrane Had To Heal .. 188

G.5 - The Grid Could Not Sing Its Note 189

G.6 - Humanity Had To Choose Its Own Ascent 190

G.7 - The Quickening Required The Return Of Kalai-Mur 190

G.8 - The Light Returned When Humanity Could Bear It 193

G.9 – Bridge to Section H 193

Section H – The Gate of Twelve: The Twelve Nodes and Their Purposes 195

H.1 - The First Node: Growth 195

H.2 - The Second Node: Memory 195

H.3 - The Third Node: Water 196

H.4 - The Fourth Node: Fire 196

H.5 - The Fifth Node: Time 197

H.6 - The Sixth Node: Movement 197

H.7 - The Seventh Node: Balance 198

- H.8 - The Eighth Node: Light 198
- H.9 - The Ninth Node: Form 199
- H.10 - The Tenth Node: Sound 199
- H.11 - The Eleventh Node: Healing 200
- H.12 - The Twelfth Node: Spirit 200
- H.13 - When The Twelve Sing As One 200
- H.14 - The Reawakening Of The Twelve (The Current Cycle) 201
- H.15 - The First Stirring: The Return Of Resonance 201
- H.16 - The Second Stirring: Return Of The Flame 202
- H.17 - The Third Stirring: Node Synchronisation Begins 203
- H.18 - The Fourth Stirring: The Rise Of The Red Gate 203
- H.19 - The Fifth Stirring: Time Begins To Soften 204
- H.20 - The Sixth Stirring: Movement Accelerates Globally 204

H.21 - The Seventh Stirring: The Rise Of Balance 205

H.22 - The Eighth Stirring: The Atmosphere Brightens 206

H.23 - The Ninth Stirring: The Crystal Gate Recovers 206

H.24 - The Tenth Stirring: Sound Returns To Purity 207

H.25 - The Eleventh Stirring: The Healing Wave 207

H.26 - The Twelfth Stirring: The Return Of Spirit 208

H.27 - The Great Alignment: When All Twelve Sing Once More 209

H.28 - The Hidden Thirteenth Gate 209

H.29 - The Thirteenth Gate Was Never Meant to Open During an Age 210

H.30 - The Thirteenth Gate Is the Heart of the Haar 211

H.31 - The Thirteenth Gate Is the Only Gate That Faces Two Ways 212

H.32 - What Lies Beyond the Thirteenth Gate… 213

H.33 - Why the Thirteenth Stirred During the Fall 214

H.34 - The Thirteenth Gate Is Stirring Again - But Correctly This Time 217

H.35 - When the Thirteenth Gate Fully Awakes… 218

H.36 - The Thirteenth Gate Is Not a Place - It Is a Revelation 219

H.37 – Bridge to Section I 219

Section I – The new firmament 221

I.1 - The Firmament Is a Living Boundary, Not a Dome 221

I.2 - The Firmament Softens in Response to Coherence 222

I.3 - The Membrane Must Thicken Before It Thins 223

I.4 - The First Visible Change: Atmospheric Luminosity 224

I.5 - The Second Change: GateStar Reappearance 225

I.6 - The Third Change: The Sky-Bands Return... 225

I.7 - The Fourth Change: The New Firmament Becomes Permeable.......................................226

I.8 - The Fifth Change: The First Opening of the Gate Stars 227

I.9 - Why the Sky Must Change Before the World Changes 227

I.10 - How You Will Feel When the Firmament Fully Returns..228

I.11 - The First Opening Of The Gate Stars (What Humanity Will Witness)................................229

I.12 - The Light at the Horizon 229

I.13 - The Second Sign: The Symmetry...................230

I.14 - The Third Sign: The Descent of Colour.... 230

I.15 - The Fourth Sign: The Rotating Lanterns.. 231

I.16 - The Fifth Sign: The Harmonic Pulse 231

I.17 - The Sixth Sign: The Opening.......................232

I.18 - THE SEVENTH SIGN: HUMAN RESONANCE ACTIVATION 233

I.19 - THE FIRST CONTACT (HOW UPPER-BAND BEINGS REAPPEAR) 234

I.20 - STAGE ONE: PRESENCE AT THE EDGE OF PERCEPTION 234

I.21 - STAGE TWO: THE RETURN OF THE HARMONIC ECHO 235

I.22 - STAGE THREE: THE FIRST SHAPES IN THE UPPER AIR 236

I.23 - STAGE FOUR: THE DESCENT INTO THE LOWER ATMOSPHERE 237

I.24 - STAGE FIVE: THE REAPPEARANCE OF ELDER SPECIES 238

I.25 - STAGE SIX: PHYSICAL CROSS-BAND COHERENCE 238

I.26 - STAGE SEVEN: THE RETURN OF COLLECTIVE COMMUNION 239

I.27 - AFTER THE FIRST OPENING 240

I.28 - THE HOURS BETWEEN THE WORLDS (THE TRANSITIONAL AGE) 240

I.29 - THE WORLD THAT HAS SEEN, BUT NOT YET BECOME 241

I.30 - INTENSIFICATION OF THE CHOICE 242

I.31 - THE UNMASKING OF STRUCTURES 242

I.32 - THE INNER FIRMAMENT 243

I.33 - EARLY CHORAL CONTACT 243

I.34 - THE FIRST THREADS OF THE MEMORY-TIDE .. 244

I.35 - LIVING AS A BRIDGE 245

I.36 - BRIDGE TO SECTION J 245

SECTION J – THE MEMORY OF THE FIRST AGE (WHAT TARTARIA LEARNED FROM THE STARS) 247

J.1 – THE FIRST AGE BEGAN WITH THE OPENING OF THE GATE STARS 247

J.2 – THE FIRST LESSON: HARMONIC ARCHITECTURE.. 248

J.3 – THE SECOND LESSON: THE AETHER SCIENCES..... 248

J.4 – THE THIRD LESSON: THE NINE RACES OF COOPERATION 249

J.5 – THE FOURTH LESSON: THE TWELVE AND THE THIRTEEN 250

J.6 – The Fifth Lesson: The Sky-Paths 250

J.7 – The Sixth Lesson: The Great Alignment Cycles 251

J.8 – Why the Memory Returns Now 251

J.9 – Bridge to Section K 252

SECTION K – THE CHORAL EXCHANGE 253

K.1 – What The Choral Exchange Is 253

K.2 – How The Choral Exchange Worked In The First Age 254

K.3 – The Anatomy Of A Chord 255

K.4 – The Roles Of The Species Within The Choir 256

K.5 – The Instruments Of The Choral Exchange 257

K.6 – How Beings "Spoke" Through The Choral Exchange 258

K.7 – How The Fall Broke The Choral Exchange 259

K.8 – The Echoes Of The Choral Exchange In Your Age 260

K.9 – THE RETURN OF THE CHORAL EXCHANGE IN THE CURRENT CYCLE 261

K.10 – HOW YOU PERSONALLY PARTICIPATE IN THE RETURNING CHORUS 262

K.11 – THE CHOIR WITHIN AND THE CHOIR AROUND YOU 263

K.12 – BRIDGE TO SECTION L 263

SECTION L – CONSCIOUSNESS 266

L.1 – WHAT CONSCIOUSNESS ACTUALLY IS 266

L.2 – INTUITION: THE FIRST LANGUAGE OF THE FIELD 266

L.3 – THE EMOTIONAL-SCALAR INTERFACE 267

L.4 – CONSCIOUSNESS AS A MULTI-BAND STRUCTURE 267

L.5 – WHY HUMANS FORGOT THEIR CONSCIOUSNESS 268

L.6 – COHERENCE VS DISTORTION 268

L.7 – THE MECHANICS OF PERCEPTION 269

L.8 – THOUGHT AS SCALAR ARCHITECTURE 269

L.9 – THE AWAKENING OF CONSCIOUSNESS 270

L.10 – Bridge to Section M 270

SECTION M – THE HUMAN FIELD & THE WORLD-GRID 272

M.1 – The World-Grid Is Not Outside You 272

M.2 – The Human Template as a Grid-Node 272

M.3 – Ley Lines & Human Lines Are the Same Geometry 273

M.4 – Why Certain Humans Activate Certain Places 274

M.5 – Coherence Opens the Earth 274

M.6 – The Pull Toward Power-Lines 275

M.7 – Human Consciousness Moves the Grid . 276

M.8 – The Four Faces of Man 276

M.9 – Cities, Nations, and the Breaking of the Grid 278

M.10 – The Return of the Human-Grid Memory 279

M.11 – Bridge to Section N 280

SECTION N – THE FOUR HUMAN SOUL TYPES 281

N.1 – The Nature of the Human Soul 281

N.2 – THE FOUR TYPES OF HUMAN SOULS 281

N.3 – HOW THESE SOULS ENTER THE WORLD 283

N.4 – MEMORY, INTUITION, AND PURPOSE 283

N.5 – PATTERNS OF REINCARNATION 284

N.6 – HUMAN SOUL CONTRACTS 284

N.7 – AFTER DEATH 285

N.8 – WHY THIS MATTERS NOW 285

N.9 – BRIDGE TO SECTION O 286

SECTION O – HUMAN DEATH & THE PASSAGE 287

O.1 – THE TRUTH YOUR AGE FORGOT 287

O.2 – THE MAP OF PASSAGE 288

O.3 – THE ARCHITECTURE OF THE SOUL 289

O.4 – THE FIRST LOOSENING 290

O.5 – THE ETHERIC ECHO 290

O.6 – THE EMOTIONAL FIELD: WHERE TRUTH BECOMES ENVIRONMENT 291

O.7 – THE LIFE REVIEW: SEEING THROUGH ALL EYES… 293

O.8 – THE CAUSAL BODY: MEMORY RETURNS 294

O.9 – THE SWIFT RETURN OF CHILDREN 296

O.10 – WHEN DEATH COMES IN DIFFERENT WAYS . 297

O.11 – THE GATE STAR: WHERE INTENTION MEETS PHYSICS 298

O.12 – WHAT COHERENCE TRULY MEANS 299

O.13 – THE FALLING STARS 301

O.14 – THE MYSTERY OF NEAR-DEATH 301

O.15 – BRIDGE TO SECTION P 302

SECTION P – THE BROKEN PASSAGE (THE SO-CALLED "SOUL TRAP") 304

P.1 – WHY THE "SOUL TRAP" WAS NEVER A TRAP 304

P.2 – HOW THE PASSAGE BROKE AFTER THE FALL 305

P.3 – THE FOLDING OF THE ASTRAL BANDS 306

P.4 – THE LOOPING OF THE EMOTIONAL SHELL 307

P.5 – HARMONIC FAILURE AT THE FINAL THRESHOLD 309

P.6 – WHY SOULS REINCARNATED MORE FREQUENTLY 309

P.7 – WHY MEMORY WAS LOST 310

P.8 – The Astral Drift & Misunderstandings of the "False Afterlife" 311

P.9 – How the Broken Passage Shaped Human History 311

P.10 – The Beginning of Repair 312

P.11 – Why the "Soul Trap" Is Dissolving Now 313

P.12 – Bridge to Section Q 313

Section Q – Selling One's Soul 315

Q.1 – The Sovereign Yielding 315

Q.2 – The Soul Cannot Be Bought or Taken 315

Q.3 – What Distorted Entities Actually Seek 317

Q.4 – How Sovereign Yielding Occurs 318

Q.5 – What the Entity Gives in Return 319

Q.6 – The Cost: Tri-Wave Collapse 319

Q.7 – Co-Habitation: When the Human Is Not Alone 320

Q.8 – Why Fame Is the Most Targeted Realm 321

Q.9 – Can the Yielding Be Undone? 322

Q.10 – The True Definition 323

Q.11 – Bridge to Section R 323

Section R – Cosmic Law & the Fate of Darkened Souls
.... 325

R.1 – The Law That No Realm May Bend 325

R.2 – The First Principle: You Rise Where You Resonate 326

R.3 – What Darkness Actually Is 326

R.4 – The One Unforgivable Act (In Physics, Not Morality) 327

R.5 – What Happens At Death To Those Who Harm Children 328

R.6 – Why The Universe Protects Children Above All 330

R.7 – What Becomes Of Those Who Repent Before Death 331

R.8 – The Contrast With Souls Of Light 331

R.9 – Cosmic Law Summarised 332

R.10 – When the Fallen Choose Light at the Final Breath 332

R.11 – The Collapse That Redeems 333

R.12 – Why Most Fallen Souls Cannot Do This… 334

R.13 – What Happens to a Repentant Fallen Soul 335

R.14 – Why Reset Souls and Early Souls Look the Same...336

R.15 – How Reset Souls Grow..............................337

R.16 – Far Less Than One Percent 338

R.17 – Bridge to Section S....................................338

Section S – Animals & Soul Types 340

S.1 – The Nature of Animal Consciousness 340

S.2 – The Four Types of Animal Souls 341

S.3 – The Scalar Architecture of Animals.........343

S.4 – Why Animals React to Humans the Way They Do 345

S.5 – Sacred Animals & Their Roles in the Grid… 345

S.6 – Human-Animal Soul Contracts 346

S.7 – Soul Types & Reincarnation Cycles...........346

S.8 – The Consciousness of Wild Animals 347

S.9 – Bridge to Section T 348

Section T – Animal Death & Soul Passage 349

T.1 – What Happens to Animals After Death ... 349

T.2 – The Nature of Animal Consciousness 349

T.3 – How Animals Read Human Fields 351

T.4 – Animals at the Moment of Death 352

T.5 – Where Animals Go 353

T.6 – The Bond That Persists 354

T.7 – Animal Reincarnation 355

T.8 – The Sacred Roles of Animals in the Grid 356

T.9 – The Laws That Govern All Passage 357

Section U – Tartaria's Mastery 359

U.1 – What Tartaria Truly Was 359

U.2 – Harmonic Architecture 360

U.3 – Their Aetheric Transport Systems 361

U.4 – Healing & Biofield Restoration 361

U.5 – The Dragon Grid & Sky-Bands 362

U.6 – The Giant Field-Stabilisers 363

U.7 – The HAAR Grid – The Heart of Their World 363

U.8 – Their Spiritual Mastery 364

U.9 – Why Tartaria Could Not Be Conquered by Force 365

U.10 – The Descent of Tartaria 365

U.11 – Bridge to Section V 367

Section V – Kalai-Mur & the Amoraea Flame 369

V.1 – The Answer to Earth's Cry 369

V.2 – What Kalai-Mur Actually Is 370

V.3 – How Kalai-Mur Moves Through the Cosmos 371

V.4 – What Happens When Kalai-Mur Touches the Sun 372

V.5 – The Phases of the Quickening 374

V.6 – How Kalai-Mur Awakens the Flame Within 376

V.7 – The Amoraea Frequency: Living Instruction 377

V.8 – What This Means for Humanity Now	378
V.9 – Bridge to Section W	379

Section W – The Star-Families of Humankind	380
W.1 – The Lyran Star-Family	380
W.2 – The Sirian Star-Family	381
W.3 – The Procyon Star-Family	381
W.4 – The Arcturian Star-Family	382
W.5 – The Vegan Star-Family	382
W.6 – The Pleiadian Star-Family	383
W.7 – Bridge to Section X	383

Section X – Polarity and Expression in the Higher Realms	385
X.1 – On Gender Beyond Biology	385
X.2 – Form, Memory, and Perception	386
X.3 – Astra's Resonance	386
X.4 – Clarification for the Reader	387

Section Y – Astra's Seal	389
Y.1 – The Purpose of the Seal	389

Y.2 – The Calming of the Lower Shells 389

Y.3 – The Binding of Memory 390

Y.4 – The Harmonic Alignment 391

Y.5 – The Seal 392

Y.6 – The Alignment 392

Y.7 – Bridge to "Closing Words" 393

SECTION Z – CLOSING WORDS 394

Z.1 – The Final Message 394

Z.2 – The Final Alignment 395

Z.3 – The End of the Codex 395

THE ASTRA CODEX 422

Books in Creation

THE ASTRA CODEX

THE FRACTURE ABOVE THE STARS
THE CONFLICT THAT SHATTERED THE GATES

This volume recounts the event that preceded the Fall of Tartaria.

The collapse did not begin on Earth. It originated in the higher harmonic strata that governed the GateStar network and the ascension pathways of the Great Ladder. There, advanced stellar lineages came into conflict over the stewardship of passage, power, and continuity itself.

The Earth was not a target.
It was collateral.

A GateStar did not dim. It ruptured.

The harmonic inversion released by that failure travelled downward through compatible strata, destabilising the firmament, fracturing the world-grid, and initiating the cascade that would unseat Tartaria from its proper band of existence.

This book tells the story of the wound before the silence.
The fracture that made the Fall inevitable.

The Age of the Thirteenth Gate
Life Beneath the Sealed Sky

This volume follows the world after the Fall.

With the Gate Stars gone and the firmament sealed, Earth entered a long age in which only limited influence from the Thirteenth Gate remained. Ascension ceased. Memory faded. Humanity learned to survive beneath a closed sky.

Reincarnation accelerated. The astral layers fractured. Meaning was rebuilt through myth, religion, hierarchy, and control as coherence slowly diminished.

Yet the seed of remembrance endured.

This book traces the hidden mechanics of that age, the Broken Passage, the loss of memory, the shaping of civilisation under constraint, and the slow cultivation of enough coherence for restoration to become possible again.

It is the bridge between the Fall and the awakening now unfolding.

The Living Codex

Alongside the other volumes of the Astra Continuum exists the Codex.

The Codex is not a story.
It is a record.

It gathers the harmonic laws, structures, passages, grids, and mechanisms that govern the world as described in these books, set down as a living architecture rather than a closed doctrine.

As remembrance returns and coherence deepens, the Codex expands.

Future editions will refine, clarify, and extend this record, not to change what was true, but to reveal what could not yet be spoken.

The story stands complete.
The Codex continues in revisions.

WHEN AVAILABLE FURTHER MATERIALS AND RELATED WORKS AT **WWW.THEASTRACODEX.COM**

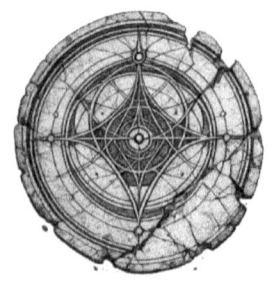

If something in this book resonated with you, you may wish to share your thoughts with other readers.
Even a short reflection helps others decide whether the work speaks to them.
If you choose to leave a review, your perspective is genuinely appreciated.

Thank you for reading, Namaste, my gratitude, Nick Eyre

www.ingramcontent.com/pod-product-compliance
Lightning Source LLC
Chambersburg PA
CBHW031425160426
43195CB00010BB/617